CW00942190

The JCT Design and Build Contract

The JCT Design and Build Contract

David Chappell
and
Vincent Powell-Smith

Second Edition

Blackwell
Science

Blackwell Science Ltd
Editorial Offices:
Osney Mead, Oxford OX2 0EL
25 John Street, London WC1N 2BL
23 Ainslie Place, Edinburgh EH3 6AJ
350 Main Street, Malden
MA 02148 5018, USA
54 University Street, Carlton
Victoria 3053, Australia
10, rue Casimir Delavigne
75006 Paris, France

Other Editorial Offices:

Blackwell Wissenschafts-Verlag GmbH
Kurfürstendamm 57
10707 Berlin, Germany

Blackwell Science KK
MG Kodenmacho Building
7-10 Kodenmacho Nihombashi
Chuo-ku, Tokyo 104, Japan

First published 1993
Second Edition published 1999

Set in 10.5/12.5pt Palatino
by DP Photosetting, Aylesbury, Bucks
Printed and bound in Great Britain by
MPG Books Ltd, Bodmin, Cornwall

DISTRIBUTORS

Marston Book Services Ltd
PO Box 269
Abingdon
Oxon OX14 4YN
(*Orders:* Tel: 01235 465500
Fax: 01235 465555)

USA
Blackwell Science, Inc.
Commerce Place
350 Main Street
Malden, MA 02148 5018
(*Orders:* Tel: 800 759 6102
781 388 8250
Fax: 781 388 8255)

Canada
Login Brothers Book Company
324 Saulteaux Crescent
Winnipeg, Manitoba R3J 3T2
(*Orders:* Tel: 204 837 2987
Fax: 204 837 3116)

Australia
Blackwell Science Pty Ltd
54 University Street
Carlton, Victoria 3053
(*Orders:* Tel: 03 9347 0300
Fax: 03 9347 5001)

A catalogue record for this title
is available from the British Library

ISBN 0-632-04899-9

Library of Congress
Cataloging-in-Publication Data
Chappell, David.
The JCT design and build contract/David Chappell and Vincent Powell-Smith. – 2nd ed.
p. cm.
Includes indexes.
ISBN 0-632-04899-9
1. Construction contracts—Great Britain.
I. Powell-Smith, Vincent. II. Title.
KD1641.C485 1999
343.41'07869—dc21 99-30890
CIP

For further information on Blackwell Science, visit our website: www.blackwell-science.com

Contents

Preface to the Second Edition

The JCT Standard Form of Building Contract With Contractor's Design has proved to be very popular. Although it is not the only design and build contract, it is certainly the best known. Design and build is still a very popular form of procurement and there is little sign that the trend is decreasing despite what some surveys suggest. The popularity of design and build owes a lot to the perception of a single point responsibility, virtually guaranteed price and time and reduced claims opportunities if certain key principles are observed. The extent to which that perception is justified will become clear.

Originally produced in 1981, the contract became known as CD 81. It has just been reprinted, incorporating all amendments, as WCD 98. We shall have to learn to call it that. The contract structure is very similar to JCT 98 and JCT 80 before that. Indeed, much of the wording is identical. This has led to problems as the parties may overlook the very many subtle, and some quite clear differences. Misunderstandings and disputes may result. It is distressingly common to find that architects acting as employer's agent deal with the contract as though it was the traditional standard form. That is a recipe, if not for disaster, at least for substantial claims. The fact is that WCD 98 is a very complex document. We suspect that many of its provisions are not, when strictly construed, what were intended by the draftsmen. Problem areas are still:

- the allocation of design responsibility
- statutory requirements
- discrepancies
- the role of the employer's agent
- payment provisions, and
- the approval of drawings.

Since the first edition, many things have changed. Now there are six more JCT amendments which together with numerous corrections have been incorporated into the reprinted WCD 98. The Housing Grants, Construction and Regeneration Act 1996, Part II, came into force on 1 May 1998 together with the Scheme for Construction Contracts (England and Wales) Regulations 1998 and resulted in the massive amendment 12 which responded to the Act by putting in place the adjudication procedure, payment and withholding of payment notice procedures and the contractor's new right to suspend work for non-payment. The amendment also contained provisions for advance payment and off-site materials bonds, interest on late payment and changes to relevant events and loss and/or expense matters. The Arbitration Act 1996 and the Construction (Design and Management) Regulations 1994 are in force and the Standard Form of Agreement for the Appointment of an Architect (SFA/99 and CE/99) are expected to be in use by the middle of the year. The amount of case law has also been increased. This edition takes account of all these changes.

This book is designed to operate on two levels: as a practical guide to assist the user in what to do next, and as an authoritative text with references to appropriate case law. Where the meaning is obscure and judicial pronouncements offer no guidance, we have taken a view and given advice appropriate to the situation. The text is illustrated, where possible, with examples of the way the contract works in use.

A common method of writing about building contracts is to provide a commentary clause by clause. In eschewing this approach, we have dealt with the contract by examining the roles of the participants and then considering particular important topics such as determination, claims and payment. For ease of reference, some of the information is also provided in tabular form.

It is hoped that the book will be useful for employers about to embark on design and build for the first time, to the contractor, to the professional acting as employer's agent, whether architect, engineer or surveyor, and to the design team acting for the contractor.

A final chapter examines the Scottish Building Contract with Contractor's Design in order to complete the usefulness of this book.

I have been responsible for this edition, but much of the text still bears many of the hallmarks of the late Professor Vincent Powell-Smith and, therefore, I have taken the liberty of continuing to refer to 'we' occasionally. I have again managed to persuade my friend and colleague, Derek Marshall LLB(Hons), MASI, ACIArb, to completely revise the chapter on dispute resolution to deal with the considerable changes which have taken place in the last few years. He has done so with his usual perceptive insights.

The male pronoun has been used throughout this book, as in the contract, for the sake of simplicity and it should be taken to mean male or female persons as appropriate.

David Chappell
C/o Chappell-Marshall Limited
27 Westgate
Tadcaster
North Yorkshire
LS24 9JB
February 1999

CHAPTER ONE
INTRODUCTION

1.1 *Definitions*

In the traditional procurement scenario, an employer appoints an architect to design a building. The architect prepares designs, seeks approval from his client and steers the project through all the stages of what is commonly known as the RIBA Plan of Work. This includes obtaining planning permission, seeking tenders, dealing with the contract and administering the contract during operations on site. Throughout, the architect acts for his client and gives him a professional service perhaps modified to suit particular client preferences. Essentially, design is in the hands of the architect who develops it into production information while construction is carried out by the contractor precisely in accordance with the architect's designs. This is still the single most popular category of procurement of buildings in this country, although within the category there are variants such as management contracting, construction management, project management, etc.

Design is a difficult concept to define. It has any number of connotations as the various dictionary definitions make plain. It can be 'a preliminary plan or sketch for the making or production of a building' as well as 'the art of producing these'. Design may be a scheme or plan of action and it can be applied equally to the work of an architect in formulating the function, construction and appearance of a building as to an engineer determining the sizes of structural members and clearly it involves the selection of materials suitable for the purpose of the proposed structure. It is generally accepted that an architect is designing not only when he produces presentation drawings showing the way the building will look, but also when he produces constructional or 'working' drawings showing how the component parts of the building fit together. He is also most certainly designing when he produces large scale details of various parts of the building and when he prepares the detailed written specification. On the other hand, the contractor is not designing when he puts the components together in a way and using materials specified by the architect. Yet the contractor may be involved in some design even in a building erected under a traditional procurement system. Consider a piece of built-in joinery designed by the architect as part of the building. He may design it in great detail and draw full-size sections through its parts, but it is still likely that he will fail to design the joinery in every detail. If there is any portion not so designed, it is possible that the contractor will assume some design responsibility if he carries on and produces what he 'knows' will be required rather than asking the architect for more information. The point is not without doubt, however, and there are cases which appear to point in different directions.

In the Australian case of *Cable (1956) Ltd* v. *Hutcherson Bros Pty Ltd* (1969), for example, although the contractor had tendered for the design, supply and installation of a bulk storage and handling plant to be built on reclaimed harbour land, the contract required the contractor's drawings to be approved by the employer's engineer. The drawings as approved showed ring foundations for storage bins. When these were erected and filled, subsidence occurred. The High Court of Australia held that, on the true interpretation of the contract documents, the contractor was not liable as he was not responsible for the suitability of the design. In the court's view, the contractor 'promised no more than to carry out the specified work in a workmanlike manner' and it would appear that the employer had not in fact relied on the contractor's skill and judgment in respect of the design.

In *Brunswick Construction Ltd* v. *Nowlan* (1974), however, Nowlan engaged an architect to design a house and then contracted with Brunswick to erect it to the architect's design. No architect supervised the construction. The design was defective and made insufficient provision for ventilating the roof space. Surprisingly, perhaps, the Supreme Court of Canada held the contractor liable for a resultant attack of dry rot, on the basis that an experienced contractor 'should have recognised the defects in the plans ... knowing the reliance which was being placed upon it'. It should have been obvious to the builder that the building would not be reasonably fit for its intended purpose if it was constructed in accordance with the defective plans.

Even if the architect remembers to draw sections through every portion, he is very unlikely to include details of the screws holding everything together. He will assume, probably correctly, that the joiner will know the kind of fixings, sizes, materials and spacing required. This is commonly referred to as 'second order design'. Architects vary in the amount of second order design they leave to the contractor and it is very difficult in some instances to decide what is the difference between this category of design and workmanship. In practice, this can lead to problems in allocating responsibility where traditional procurement paths are taken.

The idea of design and build is that the design and the construction of a project are in the hands of one firm. This appears to make eminent sense in that, in theory at any rate, it results in one point responsibility. In practice, it is not so simple. There are many terms which seem to be used indiscriminately for design and build. There are differences. The main types of design and build are as follows:

- *Design and build:* The contractor takes full responsibility for the whole of the design and construction process from initial briefing to completion of the project. This is the term which the industry tends to use as the general name for all variants of this procurement category.
- *Design and construct:* This is a wider term and it includes design and build, but also other types of construction such as purely engineering works of various kinds.
- *Develop and construct:* This is a term which lacks precision, but which is often used to describe a situation where a contractor is called upon to take a design which is partially completed and to develop it into a fully detailed design before being responsible for construction. Whether, in such a situation, the contractor is responsible for the original design as well as the development work will depend on the precise terms of the contract. However, where this type of design and build is carried out on a simple exchange of letters, it is probable the contractor is responsible for the whole of the design.
- *Package deal:* This term can be used to refer to either of the previous situations. Strictly, the term suggests that the contractor is responsible for providing everything. It particularly refers to systems of industrialised buildings which can be purchased and erected as a 'package'. The employer will usually be able to view similar completed buildings before proceeding. Closed systems of industrialised building are indicated.
- *Turnkey contract:* This is a procurement method in which the contractor really does do everything, including providing the furniture if required. The idea is that when the employer takes possession, all that remains to be done is to turn the key. It has

been said that this is not a term with a precise legal meaning: *Cable (1956) Ltd* v. *Hutcherson Bros Pty Ltd* (1969). However, another view is that the use of such a term in contract documents is likely to indicate that the contractor is undertaking at least some design responsibility.
- *Design and manage:* This is not strictly design and build at all, but simply an architect-led version of the contractor-led construction management.

It is, however, important to determine whether the contract is a traditional one or a true design and build contract because where the contractor offers not only to undertake the construction work but also to perform some or all of the design duties usually undertaken by the employer's professional team, then unless the express terms of the contract provide otherwise, the design and build contractor will be under an obligation to ensure that the building as designed is suitable for its intended purpose.

This is well illustrated by *Viking Grain Storage Ltd* v. *T.H. White Installations Ltd* (1985) where contractors undertook to design and build a grain storage and drying installation. The installation was defective. The plaintiffs alleged that some of the materials used were defective, that some of the construction work was badly performed, and that aspects of the design were unsuitable. The installation was not fit for its intended purpose. On a preliminary issue, Judge John Davies QC held that the defendants were strictly liable. The fact that they had used reasonable care and skill was no defence. There was an implied obligation that the finished installation would be fit for its intended purpose. Nothing in the express terms of the contract contradicted this obligation. The design and build contractor's liability is, in the absence of an express term to the contrary, equivalent to that of a supplier of goods, the only proviso being that the employer must have relied on his skill and judgment.

1.2 Advantages and disadvantages

The advantages of design and build are usually said to be as follows:

- The employer has a single point of responsibility to which he can refer throughout the procurement process and after construction is complete if there are any latent problems. This is in contrast to the traditional systems where the employer's point of contact is

the architect, but if there are difficulties, responsibility may lie with any one or more of a range of firms including the contractor, the architect, quantity surveyor, engineer and other consultants.
- The cost is virtually guaranteed and there is a better than average chance of meeting a fixed completion date.
- The total procurement period is likely to be shorter than a similar project using traditional methods. This is because the contractor is in charge of the whole process.
- Except when WCD 98 is used, the contractor undertakes that the finished building will be fit for its purpose.
- Buildability of the design concept.
- Likely to be fewer claims, because the factors which commonly trigger such claims are mainly under the control of the contractor.

Disadvantages are said to be as follows:

- The system is not flexible. If the employer makes any changes in his requirements, it opens the door to claims for extensions of time and direct loss and/or expense.
- The Employer's Requirements must be prepared carefully so as to accurately reflect his wishes while giving proper scope to the contractor. The contract is unforgiving to the extent that badly assembled Requirements will result in Contractor's Proposals which do not satisfy the employer.
- Because the relationship between employer and contractor's architect is not the close one of client and independent consultant, because the employer will not usually choose his own architect and because the architect may be under instructions from the contractor to design down to a price, the quality of design may not be as good as a building produced in the traditional fashion.
- The employer will be involved in additional fees. The design fees which the employer would normally pay to his consultants will be included in the contractor's design and build price. The employer will need independent professional advice and, therefore, he will have to pay extra for it.

1.3 The architect's role

Architects are said to dislike design and build. There are several reasons advanced for this, including the suggestion that where the client does not appoint an architect, the standard of design will necessarily suffer. It is difficult to understand why an architect should not be required under a design and build procurement system. Even the headnote to the JCT Standard Form of Building Contract With Contractor's Design 1998 states:

> 'This Form is not suitable for use where the Employer has appointed an Architect/a Contract Administrator to prepare or have prepared drawings, specifications and bills of quantities and to exercise during the contract the functions ascribed to the Architect/the Contract Administrator in the JCT Standard Form of Building Contract...'

Although the architect will not have the role ascribed to him under JCT 98, he cannot be discarded, because someone has to design the building. In addition, the employer will still require independent professional advice in order to use the system to best advantage. So far as the design aspect is concerned, design and build can be extremely flexible.

In order to fully understand the extent of the flexibility, it is useful to consider the extreme situations. At one extreme, the employer may approach a design and build contractor as soon as the intention to build starts to take shape. The contractor, either by means of an in-house architectural department or more commonly by sub-letting the work to an independent architect, takes details of the brief and proceeds through the stages from inception to completion. This is true design and build where the contractor is responsible for everything from start to finish. It is usual to negotiate the contractor's price, because tendering among a number of contractors is not practicable in this instance.

At the other extreme, the employer may engage a full team of consultants to act in the traditional way in taking the brief, and preparing a feasibility report, outline proposals and a detailed design together with a very full specification. Tendering then takes place and the successful tenderer proceeds on the basis that he takes responsibility for completing the detailed design as well as constructing the building. In practice, that will involve the contractor in producing a full set of production information. The employer has little to gain by adopting this system, because once the design team has designed the building, there is every reason for retaining them to deal with inspections and queries during the construction period under a traditional contractual arrangement. In any event, the employer will require some kind of independent advice at this time. One comes across this particular variant quite often and, in some instances, it originates in the employer's intention to proceed down the traditional path and its implementation until rather late in the process. The employer tries to have the best of both worlds – full control over design but with one point responsibility. In our experience, such late changes of mind result in complex design liabilities and many opportunities for claims. In such instances, the employer has no one to blame but himself and perhaps his professional advisors.

A common practice somewhere between the two extremes involves the design team in taking the brief, carrying out feasibility studies and preparing outline proposals and a performance specification for tendering purposes. The successful tenderer is responsible for completing the design using his own team and the employer's team is available to assist the employer with advice throughout the construction period. Some of these variants are shown in diagrammatic form in Fig 1.1.

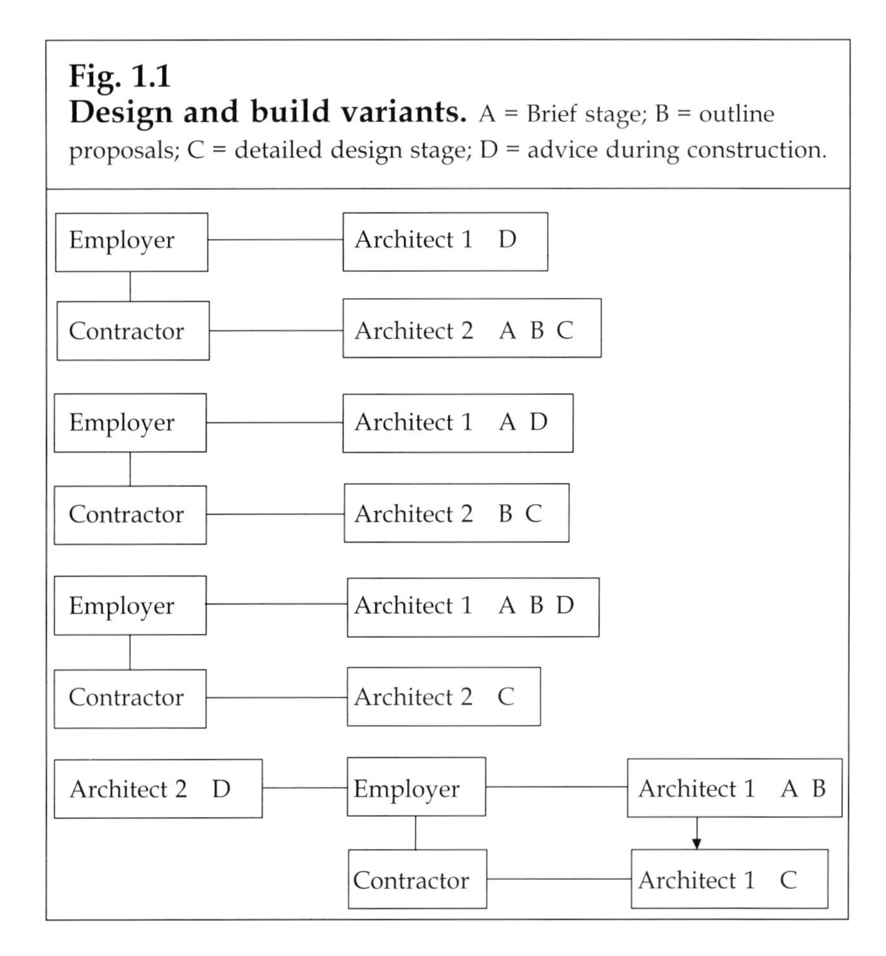

Fig. 1.1
Design and build variants. A = Brief stage; B = outline proposals; C = detailed design stage; D = advice during construction.

An interesting variation is known as consultant switch. The system requires the employer to appoint a design team in the traditional way and the team takes the employer's brief, prepares feasibility studies, and develops proposals to a fairly advanced stage with a performance specification. Tendering takes place on the basis that the successful contractor will enter into a new contract with each of the design consultants. This is supposed to avoid any danger of a design responsibility split between the employer's and the contractor's architects and it is also supposed to ensure a high degree of design continuity. Care must be taken, however, because the design team's duties to the contractor will not be the same as their duties to the employer. Therefore, simply to novate the contracts between design team and employer to the contractor will not work – although it is commonly done. The individual members of the design team and the contractor must have in their contracts with the employer the terms of the contracts between the design team and the contractor together with an undertaking to enter into a future contract on those terms. (For a detailed description of novation see Sections 3.4 and 6.1.)

Obviously, once the team have entered into contracts with the contractor, they can no longer give independent advice to the employer. With this system, the employer must either accept that he will receive no independent advice after tender stage or, and this is more likely, he will engage other consultants as necessary to provide the required advice. Problems may arise, for example if the second architect disagrees with the first architect's design. The architect's design liability in this and other situations is discussed in Chapter 3.

Another important way in which design responsibility can be split is when everything is carried out traditionally, but some element of the building, such as the foundations or a floor, is left for the contractor to design. In general, this is treated as though the element was a miniature design and build contract within the traditional contract framework. If it is thought essential to split off part of the design responsibility in this way, it is crucial that the element is as self-contained as possible, otherwise the task of sorting out respective design responsibilities becomes a nightmare.

1.4 *Standard forms available*

A multitude of forms of contract have been used for design and build procurement. It is still all too common to see traditional forms such as JCT 98 heavily amended in an attempt, rarely successful, to produce something suitable for design and build. JCT 98 is wholly unsuitable for use as a design and build contract. For many years, much design and build was carried out using contractors' in-house forms, and this is sometimes the case even today, despite the availability of an acceptable standard form of contract.

1.4.1 The Standard Form of Building Contract With Contractor's Design 1998 Edition (WCD 98)

This form was originally published in 1981 (as CD 81) by the Joint Contracts Tribunal (now the Joint Contracts Tribunal Limited) and imposes on the contractor a liability for the design equivalent to that imposed on an architect or other professional designer, i.e. an obligation to use reasonable skill and care in the preparation of the design.

CD 81 gained steadily in popularity and it became the single most commonly used form. It was modelled on the layout and wording of the Standard Form of Building Contract 1980 (JCT 80, now JCT 98), a fact which could be useful for those who were already conversant with JCT 80. There was the problem, however, that the user might not appreciate the degree to which CD 81 differed from JCT 80, often in quite subtle ways but to a significant extent. These differences persist between JCT 98 and WCD 98.

Two useful practice notes are issued by the JCT, but be aware of two words of warning. Although the practice notes are helpful in understanding what the JCT intended the contract to mean, the notes have no contractual significance themselves. A court would be interested in discovering the intentions of the parties to the contract as revealed by the words of the contract, but not any intentions on the part of the drafting body. The second warning is that the practice notes were written with reference to the 1981 edition of this contract. We understand that the notes are to be updated for the 1998 edition, but at the time of writing the contract only is available and not the notes. That is not to say that the original notes are without value. Much of what they contain is still useful:

- *Practice Note CD/1A:* This note contains an introduction, an explanation of the Employer's Requirements, the Contractor's Proposals and the Contract Sum Analysis, VAT and the contractor's design liability insurance.
- *Practice Note CD/1B:* This note contains a commentary on the

form, notes on the Contract Sum Analysis and application of formula adjustment.

The contract is intended for use where, for an agreed lump sum, the contractor will prepare and complete the design and construct the Works so as to comply with requirements stated by the employer. CD/1A states that the form is applicable where the employer has issued a document stating the employer's requirements and in response the contractor has submitted proposals for the design and execution of the Works and tendered a lump sum price.

In December 1998, the form was reprinted to incorporate amendments issued to that date. Various other changes and corrections have also been made in this edition. At the time of publication of WCD 98, the JCT had issued 12 amendments and it is probably useful to summarise them, because they are incorporated into the 1998 edition of the form:

Amendment 1 issued November 1986

This amendment dealt with partial possession and insurance matters and it substantially reproduced Amendment 2 to JCT 80. A detailed explanation of the changes is set out in Practice Note 22.

Amendment 2 issued July 1987

This amendment covered the following matters:
Arbitration.
Base date.
Discrepancies within the Employer's Requirements.
Limits to use of documents.
Statutory requirements.
Addition or deduction from the contract sum.
Kinds and standards of materials and workmanship.
Samples.
Access to site.
Imposition of restrictions.
Defects liability.
Practical completion.
Assignment of right to bring proceedings.
Sub-letting.
Insurance – personal injury or death; injury or damage to property.
Insurance – existing structures.
Deferment of possession.

Period for fixing new completion date.
Period for review of extensions of time.
Determination by employer.
Contractor's insolvency.
Corruption.
Determination by contractor.
Determination by either party.
Restoration, repair or replacement in connection with the payment provisions.
Final account.
Contribution, levy and tax fluctuations.
Off-site materials or goods.

Amendment 3 issued February 1988

This amendment dealt with the optional supplementary provisions S1 to S7 and the powers of the employer after the discovery of defective work.

Amendment 4 issued July 1988

This amendment dealt with the omission of the fair wages clauses, the conclusivity of the final account and final statement and payment period, and alterations to the arbitration clause including the introduction of the JCT Arbitration Rules 1988.

Amendment 5 issued April 1989

This amendment made changes to the VAT provisions.

Amendment 6 issued November 1990

This amendment dealt with changes in definition of the Works, introduction of 'site' for 'Works' in certain instances, provisions regarding the carrying out of work in a workmanlike manner, contractor's responsibility for subcontractors, insurance of existing structures, damages for non-completion and named persons.

Amendment 7 issued January 1994

This amendment mainly dealt with the redrafting of the determination provisions and the introduction of an optional sectional completion supplement.

Amendment TC/94/WCD

This amendment amended the insurance provisions in respect of terrorism cover and it has not been incorporated into the 1998 reprint, but it is supplied separately as an insert.

Amendment 8 issued March 1995

This amendment dealt with the incorporation of provisions relating to the Construction (Design and Management) (CDM) Regulations 1994.

Amendment 9 issued July 1995

This amendment amended the conclusivity clause referring to the final account and final statement.

Amendment 10 issued July 1996

This amendment made changes to the insurance provisions.

Amendment 11 issued May 1997

This amendment introduced clauses dealing with the Joint Fire Code.

Amendment 12 issued April 1998

This amendment introduced a substantial number of changes arising out of the proposals in *Constructing the Team* (The Latham Report) and in order to comply with the Housing Grants, Construction and Regeneration Act 1996 Part II Construction Contracts.

The WCD 98 form of contract is examined in detail in the remainder of this book.

1.4.2 The ACA Form of Building Agreement 1982, second edition 1984 (revised 1990 and 1998)(ACA 2)

This form was first published by the Association of Consultant Architects in 1982. It was subjected to much criticism – a great deal of which was emotional and unjustified – but it was amended substantially in 1984. It was again revised in minor respects in 1990 and most recently in 1998 to comply with the Housing Grants, Construction and Regeneration Act 1996 Part II. A useful *Guide to the ACA Form of Building Agreement Second Edition* is available. There is an edition of ACA 2 which has been specially adapted to the needs of the British Property Federation (BPF) system of building procurement. An important feature of ACA 2 is the provision of standard alternative clauses. Although ACA 2 is basically a traditional form, the proper combination of alternative clauses can produce a design and build variant. It is a relatively simple form with clearly defined divisions of responsibility. It is not a negotiated form like WCD 98; in the case of a dispute, any ambiguity is likely to be construed against the employer who puts it forward. The form may also be classed as the employer's written standard terms of business for the purposes of the Unfair Contract Terms Act 1977. This can affect any clauses which are deemed to be exclusions or restrictions of liability.

The key clause is clause 3.1 in which the contractor warrants that the Works will comply with any performance specification in the contract documents, and that the parts of the Works to be designed by the contractor will be fit for purpose. The design warranty could scarcely be wider and equates the contractor's position with the duty of a seller of goods to supply goods which are reasonably fit for their intended purpose. The contractor is responsible for any failure in the design irrespective of fault and the contractor must maintain design indemnity insurance under clause 6.6. The form can also accommodate small parcels of design by the contractor within a basically architect-designed framework.

Although this form has many virtues, it has not made the impact it deserves, perhaps because some contractors see it as heavily weighted in favour of the employer.

1.4.3 The Engineering and Construction Contract, November 1995, second edition (NEC)

This form was first published for the Institution of Civil Engineers in 1993. It employs a rather different philosophy to other standard forms and partly as a result, it has received some criticism. It was also the subject of fairly unrestrained praise by Sir Michael Latham in his report *Constructing the Team* (The Latham Report). It is not a specialist design and build form, but it is said to be flexible enough to support design and build as an option, somewhat like the ACA form. The basic principle of this form, which is good, is that there are a number of unchanging core clauses onto which can be grafted any one of six main option clauses (such things as priced contract with activity schedule or cost reimbursable contract, etc.). Other clauses (performance bond, retention, trust fund, etc.) can be added if desired. The contract has a strange numbering system (e.g. 40.6 is a sub-clause of clause 4) and it is written in the present tense so that it is impossible to be sure which of the actions are intended to be duties and which are optional (powers). An added difficulty is that the authors appear to have eschewed any words which have ever been defined in the courts. Therefore, it is often difficult to be sure what certain words really mean. This is certainly a brave attempt to break the mould, but we cannot recommend it. A very perceptive discussion of this form was written by Donald Valentine and published in *Construction Law Journal*, 1996, vol 12, p.305.

An addendum was issued in April 1998 to take into account the Housing Grants, Construction and Regeneration Act 1996.

1.4.4 JCT Designed Portion Supplement

This supplement effectively reproduces the important provisions of WCD 98 for a small portion of a contract generally being carried out under traditional contracting procedures. It is intended for the situation when the employer wishes part of the project to be designed by the contractor. It is intended for use with JCT 80 With Quantities.

An interesting question arises concerning the architect's obligation to integrate the contractor's designed portion with the rest of the work (clause 2.1.3). The question is so common that it is worth dealing with it here since, so far as we are aware, there is no book devoted entirely to this supplement. The position is that the contractor is responsible for the integration of the design (contained in the designed portion supplement) with the rest of the Works so far as they can be ascertained by the contractor from the information supplied to him at the date of the contract. If the architect makes no further changes in the Works, the contractor must ensure that his design is properly co-ordinated with the Works as a whole. If, however, the architect issues instructions which change the requirements on which the contractor's design is based, or which change the design of the Works, it is for the architect to give such instructions as may be necessary to achieve a proper integration of either the changed contractor design into the unamended rest of the Works or the contractor design into the changed rest of the Works.

A comparison of WCD 98 and ACA 2 clauses is given in Fig 1.2.

Fig. 1.2
Comparison of clauses in standard forms of contract WCD 98 and ACA2.

Description	WCD	ACA2
Interpretation, definitions	1	23
Definitions	1.3	23.2
Notices	1.5	23.1
Period of days	1.6	
Applicable law	1.7	25
Electronic data interchange	1.8	
Contractor's general obligations	2	1
Execution of Works	2	1.1
Priority of documents	2.2	1.3
Ambiguities and discrepancies	2.3	1.5
Contractor's skill and care	2.5	1.2
Statutory requirements	6	1.6
CDM regulations	6A	26
Bills of quantities	S5	1.4
Instructions	4	8
Instructions	4.1.1	8.1
Compliance	4.1.2	
Empowering provisions	4.2	8.4
Oral instructions	4.3	8.3
Drawings, details and information	5	2 Alternative 1
Copies of contract documents	5.2	2.1
Submission of contractor's drawings, etc.	5.3	2.2
Examination of contractor's drawings, etc.	S2	2.3
Comments on drawings	S2.2	2.4
Supply of as-built drawings	5.5	
Confidentiality	5.6	3.3
Errors in drawings	2.4	3.1
Samples	8.6	3.5
Copyright	9	

Fig. 1.2 *Contd*

Description	WCD	ACA2
Work, materials and goods	8	
Workmanship and materials	8.1	
Vouchers	8.2	
Opening up and testing	8.3	
Defective work	8.4	
Failure to carry out work in a workmanlike manner	8.5	
Supervision by the contractor	10	5
Person-in-charge	10	5.2
Site manager	S3	5.2
Visits to the Works	11	4
Access	11	4.1
Visits	11	4.2
Changes and their valuation	12	17
Definition	12.1	8.2
Change instructions	12.2	8.1
Provisional sums	12.3	16.6
Valuation: price statement	12.4 Alternative A	17.1
Valuation: rules	Alternative B 12.5	17.5
Value added tax	14	16.7
Contract sum exclusive of VAT	14.2	
Possible exemption	14.3	
Vesting of property, contractor's indemnity and insurance	15	6
On-site materials	15.1	6.1
Off-site materials	15.2	6.1
Passing of risk	15	6.2
Contractor's indemnity	20	6.3
Insurance	21	6.3
Insurance of the Works by contractor (new buildings)	22A	6.4 Alternative 1
Insurance of the Works by employer (new Works)	22B	6.4 Alternative 2
Insurance of the Works by employer (existing building)	22C	6.4 Alternative 2
Insurance against collapse, subsidence, etc.	21.2	6.5

Fig. 1.2 *Contd*

Description	WCD	ACA2
Design indemnity insurance		6.6
Insurance for loss of liquidated damages	22D	
Joint Fire Code	22FC	
Practical completion and defects liability	16	12
Practical completion	16.1	12.1
Defects liability period	16.2	12.2
Instructions about defective work	16.3	12.3
Remedy for failure to make good		12.4
Completion of making good	16.4	
Partial possession	17	13
Employer may take possession	17.1	13.1
Relevant date	17.1.1	13.2
Defects	17.1.2	13.2
Insurance	17.1.3	13.2
Reduced damages	17.1.4	13.3
Assignment and sub-letting	18	9
Assignment	18.1	9.1
Sub-letting	18.2	9.2
Required sub-contract conditions	18.3	
Named sub-contractors	S4	9.3
Instructions requiring sub-letting	S4.2.2	9.5
Determination of sub-contracts	S4.4	9.7
Possession and completion	23	11
Commencement and completion	23.1.1	11.1
Deferment of possession	23.1.2	
Postponement	23.2	11.8
Use or occupation by the employer	23.3	
Damages for non-completion	24	11.3
Certificate that Works not complete	24.1	11.2
Notices prior to deduction or payment	23.2.1	11.3
Adjustment of damages	24.2.2	11.4
Extension of time	25	11.6
Notices of delay	25.2	
Fixing a new date	25.3	11.6
Review after completion	25.3.3	11.7

Fig. 1.2 *Contd*

Description	WCD	ACA2
Relevant events	25.4	11.5
Loss and/or expense	26	7
Application	26.1	7.2
Grounds	26.2	7.1
Estimates	S7.2, S7.3	7.3
Agreement of estimates	S7.4	7.4
Determination		
By employer	27	20.1
By contractor	28	20.2
By either party	28A	21
Consequences of determination		
If employer determines	27.6, .7	22.1
If contractor determines	28.4	22.2
If determination by either party	28A.2, .3, .4, .5, .6, .7	22.3
Common law rights	27.8, 28.5	22.5
Employer's licensees	29	10
Information provided in contract	29.1	10.1
Information not provided	29.2	10.2
Payment	30	16
Interim payments	30.1	16.3
Alternative A: stage payments	30.2A	
Alternative B: periodic payments	30.2B	
Applications	30.3	16.1
Interest on late payment	30.3.7	
Contractor's right to suspend	30.3.8	
Retention	30.4	16.4
Final account	30.5	19.1
Final payment	30.6	19.3
Conclusivity	30.8	
Statutory tax deduction scheme	31	24
Definitions	31.1	
Whether employer is a 'contractor'	31.2	24.1
Provision of evidence	31.3	24.2
Certificate	31.4	24.3
Vouchers	31.5	
Calculation of deduction	31.6	24.4
Errors	31.7	

Fig. 1.2 *Contd*

Description	WCD	ACA2
Fluctuations	35, 36, 37, 38	18
Contribution, levy and tax	36	
Labour and materials	37	
Price adjustment formula	38	18.1
Dispute resolution	39	25
Conciliation		25A
Adjudication	39A	25B
Arbitration	39B	25D
Legal proceedings	39C	25C

1.5 *Tendering procedures*

Although the employer may use any method he chooses to invite and accept tenders, it is wise to follow an established procedure if the employer is to obtain the right contractor providing the right building at the right price. This is true of tendering for any kind of building procurement system, but it is especially true where design and build is concerned. The employer can be vulnerable if the documentation and procedures are not properly completed and the tenderers can be put to much abortive work. It is unfortunately frequent to find that employers who put little effort into the preliminary stages are faced with additional costs during the construction period in order to get what they actually want. Design and build is not a way to avoid making important decisions about building. The success of the finished project will reflect the amount of time the employer is willing to devote to it before tender stage. The National Joint Consultative Committee for Building (NJCC) has produced a *Code of Procedure for Selective Tendering for Design and Build 1994*. It will repay careful study. Indeed, it is required reading for anyone about to embark on design and build, whether employer, contractor or employer's agent. Some of the important points are summarised below. The Code allows tenders on the basis of single stage or two stage procedures. Tenders over the currently specified value in the public sector must be invited in accordance with the appropriate EEC directives. The use of CD 81 is assumed, although the principles are said to be applicable when other forms are used. The number of firms invited to tender should be severely restricted, probably to three or four, depending on the type and size of building. It should be remembered that tendering for design and build work involves all tenderers in high cost. For this reason, the list of possible tenderers should be prepared with care. The following must be borne in mind:

- The firm's financial standing
- Recent experience of designing and building the same kind of building
- Whether design will be in-house, and if not, by whom
- General experience and reputation
- Adequacy of management
- Adequacy of capacity.

Each firm on the short list should be sent a preliminary enquiry to discover if he is willing to tender. The enquiry should contain:

- Job title
- Type of building
- Employer
- Employer's agent
- Location of site including plan
- Availability of and restrictions on services
- General description of work
- Approximate cost range
- Number of tenderers proposed
- Tendering method: single stage or two stage
- Contractor's involvement in planning procedures
- Whether a conservation area
- Form of contract and amendments
- Variable contract details
- Whether Defective Premises Act 1972 applies
- Any limit on the contractor's liability
- Nature and extent of contractor's design input
- Simple contract or contract as deed
- Anticipated date for possession
- Contract period
- Anticipated date for dispatch of tender documents
- Length of tender period
- Length of time tender must remain open for acceptance
- Liquidated damages
- Bond
- Special conditions

- Consideration of alternative tenders
- Basis of assessment of tenders.

It is of great importance that the preliminary enquiry states to what extent the acceptance will depend on factors other than price. To aid in assessment, the employer must state exactly what he requires to be submitted with the contractor's tender. Contractors who respond positively should be interviewed to reduce the choice to the pre-determined number of tenderers. If any prospective tenderer has to withdraw, he should give notice before the issue of tender documents and certainly not more than two days thereafter. Note:

- The tender documents should be despatched on the stated date
- All tenders should be submitted on the same basis
- The tender period will be not less than one month and may be several months
- The employer should consider the scope for alternative offers.

Tenderers wanting clarification should notify the employer who must inform all tenderers of his decision. Under English law, a tender may be withdrawn at any time before it is accepted. Unsuccessful tenderers should be informed as soon as a tender has been accepted or a tenderer has been selected to proceed to the second stage as appropriate. If errors are found in the priced document, the employer must take the appropriate steps as set out in the invitation to tender; either:

- Give the contractor the opportunity to confirm or withdraw his tender; *or*
- Give the contractor the opportunity to confirm his tender or to amend it to correct any genuine errors. If the contractor amends and he is then no longer acceptable, the next preferred tender should be examined.

Where single stage tendering is involved, there is still scope for negotiation if the preferred tender is too high. The Code recommends acceptance of a first stage tender, where two stage tendering is adopted, and the clear definition of the following matters:

- Grounds for withdrawal from the second stage
- Entitlement to costs and method of ascertainment if second stage negotiations are not concluded to mutual satisfaction

- Reimbursement for any work done on site if second stage procedures are abortive.

Acceptance of a first stage tender has its difficulties. It is likely that there is no true acceptance at that stage, because of the need to leave open the second stage negotiation procedure, but neither party wishes to have a concluded contract at that stage. It is probably better to avoid anything that appears to be an acceptance and the employer should simply notify the successful tenderer of his intention to continue negotiations in the hope of achieving a mutually satisfactory outcome. The second stage is really the finalisation of the contractor's proposals. Where contractors have been notified that specific conditions will apply to the tendering process, the employer must strictly adhere to such conditions. Any failure in this respect might entitle the contractor to recover damages, unless the employer has protected himself against liability by means of a suitably worded clause in the tender documentation. By setting out terms for tendering, the employer is making an offer, in a limited way, that if a tenderer submits a tender in response, the employer will proceed according to such terms. A contract is formed and breach of its terms will enable the other party to recover damages for proven loss. Such damages would generally amount to the cost of preparing the tender, which could be a substantial sum where design and build is concerned. In appropriate circumstances, it is conceivable (but debatable) that a tenderer whose tender was not properly considered could claim the loss of the profit he would have made had he been properly awarded the contract.

These propositions derive from the recent Court of Appeal decision in *Blackpool & Fylde Aero Club* v. *Blackpool Borough Council* (1990), where the defendants invited tenders for a concession. The tender document stated that the defendants did not bind themselves to accept 'all or any part of any tender' and also that 'no tender which is received after the last date and time specified will be admitted for consideration'. Tenders had to be received by the Council 'not later than 12 o'clock noon on Thursday 17 March 1983'. At 11 AM on 17 March the plaintiffs' representative put their tender into a letterbox at the Town Hall. A notice on the box stated that the box was emptied daily at 12 o'clock noon. In the event, the plaintiffs' tender was not taken from the box until 18 March, and was excluded on the ground that it was too late. The concession was awarded to another tender and the plaintiffs sued alleging breach of contract. The Court of Appeal upheld the claim, holding that there was a contractual obligation to consider any tender properly submitted in accordance with the stipulated and detailed conditions of tendering. In effect, a contract was implied.

However, the *Blackpool* case was distinguished by a differently constituted Court of Appeal in *Fairclough Building Ltd* v. *Port Talbot Borough Council* (1992) where it was held that, under the normal tendering process, a tenderer has no cause of action where his tender is rejected but has been given *some* consideration and the recipient of the tender has acted reasonably. In that case, the Council decided to have a new Civic Centre constructed, and Fairclough applied to be included on the selective tendering list. Their application was successful and subsequently they were invited to tender. The wife of one of Fairclough's directors (whose name was on the company's letterheads) was employed as an architect by the Council and very properly disclosed her 'interest' under the Local Government Act 1982. In fact the relationship was already known to the Borough Engineer, but in the result, Fairclough were removed from the tender list, although it was said that the 'decision is not intended to reflect any doubts whatsoever upon the integrity of your company or the individual member of staff'.

Fairclough considered that there was a breach of contract and, on appeal, relied on the *Blackpool* case. The Court of Appeal ruled against Fairclough and held that the Council had fulfilled its obligation by giving some consideration to the tender and had acted reasonably. *Blackpool and Fylde Aero Club* v. *Blackpool Borough Council* (1990) was distinguished on somewhat slender grounds. This decision, however, does not affect what is set out above.

CHAPTER TWO
CONTRACT DOCUMENTS

2.1 *The documents*

The contract documents can be any documents which are evidence of the contract. They are agreed by the parties to the contract and signed. It is important that each document is signed by both parties and dated. To avoid any doubt, it is customary for each document to be endorsed: 'This is one of the contract documents referred to in the Agreement dated . . .'. WCD 98, unlike some other JCT contracts, does not expressly define the contract documents. This is probably because the nature of this system of procurement precludes hard and fast rules. However, it is clear that certain categories of document must be provided. These are referred to in the recitals as:

- Employer's Requirements
- Contractor's Proposals
- Contract Sum Analysis.

In addition, contract form WCD 98 must be completed. The contents of WCD 98 are arranged as follows:

Articles of Agreement

Conditions

1 Interpretation, definitions, etc.
2 Contractor's obligations
3 Contract sum - additions or deductions - adjustment - interim payments
4 Employer's instructions
5 Custody and supply of documents
6 Statutory obligations, notices, fees and charges
6A Provisions for use where appendix 1 states that all the CDM Regulations apply
7 Site boundaries
8 Work, materials and goods

9 Copyright, royalties and patent rights
10 Person-in-charge
11 Access for employer's agent, etc. to the Works
12 Changes in the Employer's Requirements and provisional sums
13 Contract sum
14 Value added tax – supplemental provisions
15 Materials and goods unfixed or off-site
16 Practical completion and defects liability period
17 Partial possession by employer
18 Assignment and subcontracts
19 [Number not used]
20 Injury to persons and property and indemnity to employer
21 Insurance against injury to persons and property
22 Insurance of the Works
23 Date of possession, completion and postponement
24 Damages for non-completion
25 Extension of time
26 Loss and expense caused by matters affecting regular progress of the Works
27 Determination by employer
28 Determination by contractor
28A Determination by employer or contractor
29 Execution of work not forming part of contract
30 Payments
31 Statutory tax deduction scheme
32 [Number not used]
33 [Number not used]
34 Antiquities
35 Fluctuations
36 Contribution, levy and tax fluctuations
37 Labour and materials cost and tax fluctuations
38 Use of price adjustment formulae
39 Settlement of disputes – adjudication – arbitration – legal proceedings

Code of practice referred to in clause 8.4.3

Supplementary Provisions
S1 [Number not used]
S2 Submission of drawings, etc. to employer
S3 Site manager
S4 Persons named as subcontractors in Employer's Requirements

S5 Bills of quantities
S6 Valuation of change instructions - direct loss and/or expense - submission of estimates by the contractor
S7 Direct loss and/or expense - submission of estimates by the contractor

Appendix 1
Annex to Appendix 1: Terms of bonds
Appendix 2: Method of payment - alternatives
Appendix 3
Supplemental Provisions (*the VAT Agreement*)
Annex 2 to the Conditions: Supplemental conditions for EDI
Modifications (sectional completion)

For the purposes of this book we have not followed the layout of the printed form, rather we have adopted what we believe to be a more logical arrangement, dealing with the form on a topic basis and making reference to appropriate clauses, wherever they may be located.

2.2 *Completing the form*

Care must be taken in completing the form. This task is normally undertaken by the employer's professional advisor, i.e, whoever is the employer's agent under the provisions of article 3. Sometimes it is considered necessary to make amendments to the clauses in the printed form. If possible, such amendments should be avoided, but if it is not possible, each amendment or deletion should be clearly made in the appropriate place on the form and each party should initial, preferably at the beginning and end of the amendment, especially where a deletion has been carried out.

Articles of agreement

The date is always left blank until the form is executed by the parties. The names and addresses of the employer and the contractor must be inserted in the space provided. Where limited companies are involved, it is sensible to insert the company registration number in brackets after the company name so that there is no possible chance of confusion in cases where companies change or even exchange names.

The first recital is important. The description of the work must be entered with care, because among other things it can affect the operation of the variation clause (clause 12). The contract sum is to be inserted in article 2 and the name of the employer's agent must be inserted in article 3 unless the employer has unwisely decided against the employment of an agent. Normally, the agent's name will be the name of a firm.

The fifth recital states that the extent of the application of the CDM Regulations, and where sectional completion is applicable, is stated in appendix 1.

Article 7.1 must be completed if the CDM Regulations apply. That will be in virtually every instance. Article 7.1 requires the insertion of the name of the planning supervisor under the Regulations if it is not to be the contractor.

Attestation

Alternative clauses are provided to enable the contract to be executed under hand or as a deed. The most important difference between the two is that the Limitation Act 1980 sets out a limitation period which is ordinarily six years for contracts under hand and twelve years where the contract is executed as a deed. The limitation period starts to run from the date at which the breach of contract occurred. For practical purposes, the latest date from which the period would run would be the date of practical completion, this being the latest date at which the contractor could correct any breach before offering the building as completed in accordance with the contract documents: *Borough Council of South Tyneside* v. *John Mowlem & Co, Stent Foundations Ltd and Solocompact SA* (1997). Contractors will doubtless opt for contracts under hand, but employers will look to extend the contractor's liability for as long a period as possible and consequently are well advised to see that the contract is entered into as a deed.

Before the Law of Property (Miscellaneous Provisions) Act 1989 and the Companies Act 1989 came into force, it used to be necessary to seal a document in order to make it into a deed. Although sealing is still possible, it is no longer necessary and all that is required in the case of a company is that the document must state on its face that it is a deed and it must be signed by two directors or a director and a company secretary. There are slightly different requirements in the case of an individual. Examples of suitable attestation clauses for a deed are shown in Fig 2.1.

Fig. 2.1
Alternative attestation clauses for a deed.

Executed as a deed by the Employer/Contractor hereinbefore
mentioned namely...........................
by affixing hereto its common seal
in the presence of

....

....

Alternatively for companies registered under the Companies Acts:

Executed as a deed by the Employer/Contractor acting by two directors/a director and its secretary (delete as appropriate) whose signatures are here subscribed:
namely...
Signed....................................Director
Signed....................................Director/Secretary
(delete as appropriate)

To execute as a deed, an individual may sign and seal with a witness to the signature or the individual may sign in the presence of a witness who attests the signature.

Conditions

Clause 1.3: Amend the reference to public holidays if different public holidays are applicable.
Clause 1.6: If the parties do not wish the proper law of the contract to be English law, appropriate amendments must be made to clause 1.6.
Clause 22.2: If it is not possible to take out insurance against the risks covered by the definition of 'All Risks Insurance', either the definition in this clause should be amended or the risks which are actually to be covered should be inserted in place of the definition.
Clause 22: Delete two of clauses 22A, 22B and 22C depending on who is to insure (see Chapter 11).
Clause 22B.2: This clause should be deleted if the employer is a local authority.
Clause 22C: Where this clause is not deleted, clause 22C.3 must be deleted if the employer is a local authority.
Clause 30.1.1.2: This clause should be deleted if the employer is a local authority.
Clause 30.1.2.2: This clause should be deleted if the employer is a local authority.

Clause 30.4: Clause 30.4.2.2 must be deleted if the employer is a local authority (but see Chapter 10 for a discussion of the implications). *Clause 30.5:* Clause 30.5.3.11 must be deleted if the employer is a local authority.

Supplementary Provisions

If it is desired that one or more of these provisions should not apply, the appropriate provisions should be deleted.

Appendix 1

It is important that the appendix is completed so as to correspond precisely with the information given to the contractor in the invitation to tender or, if that information subsequently has been varied by agreement between the parties, the varied details must be inserted. An example of a completed appendix 1 is shown in Fig 2.2.

Appendix 2

Delete either alternative A or alternative B (second and third lines) to leave the preferred alternative undeleted. If alternative A (stage payments) is chosen, the table showing the description of the stages and the corresponding cumulative values must be completed. It is important to give thought to the description of the stages. Easily identifiable stages are essential to the smooth operation of this alternative. It is quite common, although potentially disastrous, for the stages to be completed as a number of weeks, for example, four weeks, eight weeks, twelve weeks and so on until the full contract period is reached. This can lead to the contractor demanding payment for the first and other stages even if virtually no work has been carried out. The correct method of completion is in terms of portions of building completed: for example, foundations, ground floor, first floor; or block one, block two, block three.

If alternative B (periodic payments) is desired, the period between applications for payment must be inserted. If nothing is inserted, it will be one month.

Appendix 3

The three spaces are for the insertion of sufficient details to identify the documents comprising the Employer's Requirements, the

Fig. 2.2

ⓢ

Appendix 1

Clause etc.	*Subject*	
Fourth recital and 31	Statutory tax deduction scheme	Employer at Base Date ~~*is a 'contractor'~~/is not a 'contractor' for the purposes of the Act and the Regulations
Fifth recital	CDM Regulations	*All the CDM Regulations apply/ ~~Regulations 7 and 13 only of the CDM Regulations apply~~
Article 1	Supplementary Provisions	Supplementary Provisions *to apply/~~not to apply~~
Articles 6A and 6B	Dispute or difference – settlement of disputes	*Clause 39B applies *Delete if disputes are to be decided by legal proceedings and article 6B is thus to apply *See the Guidance Note to WCD 81 Amendment 12 on factors to be taken into account by the Parties considering whether disputes are to be decided by arbitration or by legal proceedings*
1·3	Base Date	4 January 1999
1·3	Date for Completion	21 February 2000
1·8	Electronic data interchange	The JCT Supplemental Provisions for EDI ~~*apply~~/do not apply If applicable the EDI Agreement to which the Supplemental Provisions refer is: *the EDI Association Standard EDI Agreement *the European Model EDI Agreement
2·5·2	Scheme approved under S.2(1) of the Defective Premises Act 1972	________
2·5·3	Limit of Contractor's liability for loss of use etc.	Clause 2·5·3 *does not apply/ ~~applies with limit of~~ £ ________ [ii]
14·2	VAT Agreement	Clause 1A of the VAT Agreement [i] ~~*applies~~/does not apply

Footnotes

*Delete as applicable.

[ii] Complete as applicable.

[i] Clause 1A can only apply where the Contractor is satisfied at the date the Contract is entered into that his output tax on all supplies to the Employer under the Contract will be at either a positive or a zero rate of tax.

On and from 1 April 1989 the supply in respect of a building designed for a 'relevant residential purpose' or for a 'relevant charitable purpose' (as defined in the legislation which gives statutory effect to the VAT changes operative from 1 April 1989) is only zero rated if the person to whom the supply is made has given to the Contractor a certificate in statutory form; see the VAT leaflet 708 revised 1989. Where a contract supply is zero rated by certificate only the person holding the certificate (usually the Contractor) may zero rate his supply.

This footnote repeats footnote [i] *for clause 14·2.*

WCD 98 — 31

Fig. 2.2 *Contd*

Clause etc.	*Subject*	
15·2·1	Listed items – uniquely identified	*For uniquely identified listed items a bond as referred to in clause 15·2·1 in respect of payment for such items is required for £ 75,000–00 (Seventy-Five Thousand Pounds) *Delete if no bond is required
15·2·2	Listed items – not uniquely identified	*For listed items that are not uniquely identified a bond as referred to in clause 15·2·2 in respect of payment for such items is required for £ 150,000–00 (One Hundred and Fifty Thousand Pound) *Delete if clause 15·2·2 does not apply
16·2, 17 and 30	Defects Liability Period (if none other stated is 6 months from day named in the Employer's statement as to Practical Completion of the Works)	12 months
18·1·2	Assignment by Employer of benefits after Practical Completion	Clause 18·1·2 *applies/~~does not apply~~
21·1·1	Insurance cover for any one occurrence or series of occurrences arising out of one event	£ 2,000,000–00 (Two Million Pounds)
21·2·1	Insurance – liability of Employer	Amount of indemnity for any one occurrence or series of occurrences arising out of one event £ 200,000–00 (Two Hundred Thousand Pounds) [jj]
22·1	Insurance of the Works – alternative clauses	*Clause 22A/~~Clause 22B/Clause 22C~~ applies (See footnote [m] to clause 22)
*22A, ~~22B·1, 22C·2~~	Percentage to cover professional fees	Twelve Per Cent
22A·3·1	Annual renewal date of insurance as supplied by Contractor	31 July 1999
22D	Insurance for Employer's loss of liquidated damages – clause 25·4·3	Insurance ~~*may be required~~/is not required
22D·2		Period of time ~~____~~

Footnotes

*Delete as applicable.

[jj] If the indemnity is to be for an aggregate amount and not for any one occurrence or series of occurrences the entry should make this clear.

82 WCD 98

Fig. 2.2 *Contd*

Clause etc.	*Subject*	
22FC·1	Joint Fire Code	The Joint Fire Code *applies/~~does not apply~~ If the Joint Fire Code is applicable, state whether the insurer under clause 22A or clause 22B or clause 22C·2 has specified that the Works are a 'Large Project': *YES/~~NO~~ (where clause 22A applies these entries are made on information supplied by the Contractor)
23·1·1	Date of Possession	1 February 1999
23·1·2, 25·4·14, 26·1	Deferment of the Date of Possession	Clause 23·1·2 *applies/~~does not apply~~ Period of deferment if it is to be less than 6 weeks is 6 weeks
24·2·1	Liquidated and ascertained damages	at the rate of £ 1,000–00 per week
28·2·2	Period of suspension (if none stated is 1 month)	One month
28A·1·1·1 to 28A·1·1·4	Period of suspension (if none stated is 3 months)	Three months
28A·1·1·5 to 28A·1·1·7	Period of suspension (if none stated is 1 month)	One month
30·1·1·2	Advance payment	Clause 30·1·1·2 ~~*applies~~/does not apply ~~If applicable:~~ the advance payment will be **£ ________ ________% of the Contract Sum and will be paid to the Contractor on ________ and will be reimbursed to the Employer in the following amount(s) and at the following time(s) ________ ________ ________ An advance payment bond ~~*is/is not required~~

Footnotes

*Delete as applicable.

**Insert either a money amount or a percentage figure and delete the other alternative.

WCD 98 83

Fig. 2.2 *Contd*

Clause etc.	*Subject*	
30·4·1·1	Retention Percentage (if less than 5 per cent) [kk]	Five Per Cent
35	Fluctuations: (if alternative required is not shown clause 36 shall apply)	clause 36 [ll] ~~clause 37~~ ~~clause 38~~
36·7 or 37·8	Percentage addition	Three Per Cent
38·1·1·1	Formula Rules	~~rule 3: Base Month~~ ~~________ 19____~~ ~~rule 3: Non-Adjustable Element [mm]~~ ~~________ (not to exceed 10%)~~ ~~rules 10 and 30(i): Part I/Part II [nn] of Section 2 of the Formula Rules is to apply~~
39A·2	Adjudication – nominator of Adjudicator (if no nominator is selected the nominator shall be the President or a Vice-President of the Royal Institute of British Architects)	President or a Vice-President or Chairman or a Vice-Chairman: *Royal Institute of British Architects ~~*Royal Institution of Chartered Surveyors~~ ~~*Construction Confederation~~ ~~*National Specialist Contractors Council~~ *Delete all but one
39B·1	Arbitration – appointor of Arbitrator (if no appointor is selected the appointor shall be the President or a Vice-President of the Chartered Institute of Arbitrators)	President or a Vice-President: ~~*Royal Institute of British Architects~~ ~~*Royal Institution of Chartered Surveyors~~ *Chartered Institute of Arbitrators *Delete all but one

Footnotes

[kk] The percentage will be 5 per cent unless a lower rate is specified here.

[ll] Delete alternatives not used.

[mm] Only applies if Employer is a local authority.

[nn] Strike out according to which method of formula adjustment (Part I – Work Category Method or Part II – Work Group Method) has been stated in the Employer's Requirements or Contractor's Proposals as the case may be.

84 WCD 98

Contractor's Proposals and the Contract Sum Analysis respectively.

2.3 *Employer's Requirements*

This is the employer's instructions to the contractor. It is the information the contractor uses to prepare his proposals and if the Employer's Requirements are wrong, the Contractor's Proposals will be wrong. Essentially, this document is a performance specification. It should specify the criteria, whereas the traditional operational specification specifies the particular way in which criteria are to be satisfied. Thus, the document may specify a particular thermal insulation value, durability, load-bearing capacity and weather tightness for a wall which the contractor can satisfy by using a number of different materials and combinations of materials. Traditionally, the actual materials and workmanship of the wall would have been specified. Although there is provision for the employer to include bills of quantities in his Requirements (supplementary provision S5), an employer who includes bills of quantities will throw away many of the advantages offered by the design and build concept.

It is sometimes thought that design and build is a soft option for the employer. If all that is required is a very simple building – a few thousand square metres of warehousing – the Employer's Requirements can be quite brief. In most cases, however, as much effort must be devoted to producing the performance specification as would be required for the traditional specification. The contractor is not responsible for the whole of the design but only for its completion (see Chapter 3). Therefore, the less information the employer provides, the greater will be the contractor's liability. Thus, if part of the Employer's Requirements consists of a set of very advanced working drawings, the contractor will need to do little but to construct the building from those drawings and the employer will know exactly what is to be provided. On the other hand, if the Employer's Requirements are very brief and the drawings are very simple sketch drawings, the employer will have little control over the end product. Put another way, the more that is left to the contractor, the greater will be his chance to save money and put forward an attractive tender figure. In practice, the employer will specify criteria together with any particular parts of the design which are mandatory upon the contractor – for example, marble in the lobby of a large hotel –

and will make clear which aspects are left to the contractor's initiative.

It is very important that the employer crystallises his requirements before executing the contract. Although provision is made in the contract for the employer to make changes in his requirements, it is by no means as easy to do this as it is in a contract such as JCT 98 and the contractor will have the right to object to many changes (see Chapter 10). An employer who considers that he might wish to make changes once the construction has begun should seriously consider using another more suitable form of contract, because apart from other considerations, he will lose many of the advantages, in terms of risk and price, offered by this form (see Chapter 1).

It is strongly advised that the employer obtains planning permission before accepting any tender. It is perfectly possible to make the contractor responsible for obtaining such permission, but it should be remembered that actually getting permission can never be guaranteed, because it depends on the planning authority. Therefore, the situation could arise that the contractor applies unsuccessfully for planning permission, or if successful, it may take several months of negotiation before it is finalised. Not only does the contract make provision for extension of time in such cases (clause 25.4.7), that is after all only reasonable), it also entitles the contractor to loss and expense (clause 26.2.2) which is also reasonable. If the works are suspended for the period named in the appendix due to delay in obtaining planning permission, the contractor is entitled to determine his employment (clause 28A.1.1.4). It is possible for the employer to specify that amendments to comply with planning requirements are not to be treated as changes in the Employer's Requirements and, therefore, to be carried out at the contractor's own cost, but there will be a hefty price to pay at tender stage.

There are two points which merit careful attention. The first point is that many of the statements made by the employer within the Employer's Requirements will be representations. The contractor will use the information in compiling his tender. Typically, this will include information about the site and ground conditions. If any of the statements of fact are incorrect, they will probably amount to misrepresentations. A misrepresentation which is one of the inducing causes of a contract and which causes loss to the innocent party may result in legal liability.

Depending on whether the misrepresentation is innocent, negligent or fraudulent, the contractor may be able to recover damages or even to put the contract at an end if he suffers some loss thereby.

The employer may not necessarily be able to avoid the consequences of a misrepresentation by including a warning to the contractor to check, or even by including a disclaimer. It may still be held to be a misrepresentation for which the contractor has a remedy in law: *Cremdean Properties Ltd and Another* v. *Nash and Others* (1977).

A misrepresentation may also amount to a collateral warranty. For example, in *Bacal Construction (Midland) Ltd* v. *Northampton Development Corporation* (1975), which involved a design and build contract, the contractor was instructed to design foundations on the basis that the soil conditions were as indicated in borehole data provided by the employer. The Court of Appeal held that there was a collateral warranty that the ground conditions would be in accordance with the hypotheses upon which Bacal had been instructed to design the foundations and that they were held entitled to damages for its breach.

The second point is that many sets of Employer's Requirements contain a provision to the effect that workmanship and/or materials are to be to the employer's approval. The result of inserting such a provision is that when the final account and final statement become conclusive as to the balance due between employer and contractor, they are also conclusive evidence that any materials or workmanship reserved for the employer's approval are to his reasonable satisfaction, subject to very limited exceptions (clause 30.8.1.1). This provision probably shuts the door on any subsequent attempt by the employer to contend that such materials or workmanship are defective. It should be noted that it does not matter whether the employer has, in fact, satisfied himself about the materials or workmanship. If nothing is reserved to the employer's approval, the pitfall is avoided. Other phrases such as 'to the employer's satisfaction' may well have the same effect. It is recognised that there will be situations in which the employer will insist on keeping final approval to himself rather than rely on any performance criteria. Such situations should be limited and the employer or his agent must make sure that the items in question are carefully inspected before practical completion and again before the final account and final statement become conclusive. The particular wording of the contract appears to avoid the situation, recently highlighted, in IFC 84 and JCT 80 where the final certificate is conclusive regarding the architect's opinion of quality and standards whether or not expressly reserved to the architect's opinion: *Colbart Ltd* v. *H. Kumar* (1992); *Crown Estates Commissioners* v. *John Mowlem & Co* (1994). In any event, in an excess of caution, JCT probably settled the matter by the issue of Amendment 9 in 1995. The following matters should always be included in the Employer's Requirements:

- Details of the site including the boundaries (unless the site is being provided by the contractor in which case clause 7 must be amended)
- Details of accommodation requirements
- Purposes for which the building is to be used
- Any other matter likely to affect the preparation of the Contractor's Proposals or his price
- Statement of functional and ancillary requirements:
    - Kind and number of buildings
    - Density and mix of dwellings and any height limitations
    - Schematic layout and/or drawings
    - Specific requirements as to finishes, etc.
- Bills of quantities in accordance with supplementary provision S5 if required
- Details of any provisional sums
- Statement of planning and other constraints, e.g. restrictive covenants, together with copies of any statutory or other permissions relating to the development
- Statement of site requirements
- The extent to which the contractor is to base his proposals on information supplied in the Employer's Requirements
- Access restrictions
- Availability of public utilities
- The method of presentation of the Contractor's Proposals:
    - Drawings, plans, sections, elevations, details, scales
    - Any special requirements, for example, models, computer animation, video
    - Layout of specialist systems
    - Specification requirements
- If supplementary provision S2 is to be used, the employer's requirements regarding submission of contractor's drawings
- If supplementary provision S3 is used, the employer's requirements regarding the records the site manager is required to keep
- Detailed requirements in respect of the as-built drawings which the contractor must supply in accordance with clause 5.5.
- Whether stage or periodic payments are to be made
- Functions to be carried out by the employer's agent and, if required, the quantity surveyor and the clerk of works

- Information to be included for the contract:
  - The form of the Contract Sum Analysis and its content
  - Whether the employer is a 'contractor' under the statutory tax deduction scheme
  - The nominator of the adjudicator
  - If arbitration or litigation is to apply and if arbitration, the appointor of the arbitrator
  - The method of fixing the date for completion
  - The base date
  - If dwellings, whether subject to the NHBC scheme
  - Whether and to what extent there is any limit on the contractor's liability for consequential loss
  - The detailed manner in which appendix 1 is to be completed
  - List of materials for clause 15.2.1 and whether a bond is required
  - Whether advanced payment will be made and whether a bond is required.

Supplementary provision S5 sets out certain rules if the Works are described in the Employer's Requirements by bills of quantities:

- The method of measurement must be stated
- Errors in the bills must be corrected by the employer and the correction is to treated as if it were a change in the Employer's Requirements
- If a valuation is carried out under the terms of clause 12.5, rates and prices in the bills of quantities must be substituted for the reference to values in the Contract Sum Analysis
- If price adjustment formulae are to be used (clause 38), the rates and prices in the bills of quantities are to be used so far as is relevant.

For the employer to include bills of quantities in his Requirements indicates that the design of the building is very advanced. If that is the case, the amount of design left to the contractor will be very small. In that situation, it may well be advisable for the employer to continue with the project on a traditional basis. The employer could reasonably ask the contractor to provide bills of quantities as part of the Contract Sum Analysis, but they seem to have no logical place in the Requirements.

2.4 *Contractor's Proposals*

Put simply, the Contractor's Proposals should answer the Employer's Requirements. If the Employer's Requirements are detailed, the Proposals will be similarly detailed. If the Requirements are rather vague, the Proposals may well leave many loose ends and there are likely to be elements of the building which are not quite what the employer expected. Therefore, to take an extreme case, if the employer simply asked to be provided with 30,000 square metres of office space on a particular site, it will leave the contractor with tremendous scope in design, construction and costing.

Most contractors will submit a detailed specification covering all the work and materials they will use to complete the project. They may also include a programme and a method statement. It is not usual to make either of these documents a contract document, because to do so requires both employer and contractor to comply with it in every particular. If it becomes necessary for the contractor to carry out the work in a different way, he may be entitled to claim payment: *Yorkshire Water Authority* v. *Sir Alfred McAlpine and Son (Northern) Ltd* (1985).

The contractor must plug any gaps in the Employer's Requirements by including the information in his Proposals. This is particularly true about the contract data. If an important point such as the system of payment has been omitted from both documents, there is a ready-made source of dispute before the contract is executed.

It sometimes happens that the contractor wishes to propose a material or constructional detail which is contrary to what is contained in the Employer's Requirements. The two documents must be consistent and, therefore, the contractor must draw the employer's attention specifically to such a proposal so that, if accepted, the Requirements can be amended before the contract documents are signed. The contractor is best advised to make such a proposal as an alternative and subject to a stated price adjustment.

The Contractor's Proposals should not contain any provisional sums unless they are in the Employer's Requirements. If the contractor feels that a provisional sum must be included, although not requested by the employer, the employer's attention again must be drawn to the sum so that it can be included in the Employer's Requirements document before signing. The consequences of discrepancies are discussed in section 2.7.

2.5 *Contract Sum Analysis*

WCD 98 is a lump sum contract. This means that, essentially, the contractor carries out the work for a fixed and stated amount of money payable by the employer. There is no provision for remeasurement although there is provision for changes in the Employer's Requirements and fluctuations. Payment may be made by fixed stages or by periodic payment based on the value of work done. The purpose of the Contract Sum Analysis is to assist in valuation of changes and work carried out, where appropriate, and to enable fluctuations to be calculated. The employer may require the Analysis in any form and the contractor must comply. Where formula fluctuations are to be used, the Analysis must contain the appropriate information, properly arranged. Guidance can be sought from Practice Note 23 1987. Whether or not the employer so requires, the contractor should always include a method of valuing design work. This might very likely be on an hourly basis and it will be needed in the valuation of changes and also in the valuation of design work carried out, but later aborted. This is a common occurrence in design and build where the employer may ask the contractor to suggest alternative designs for part of the building, but eventually proceed with the original design on which the contractor's price was based. In the absence of a clearly laid down system of charging for such work, the contractor may find that he recovers nothing or, at best, a nominal amount.

2.6 *Supplementary provisions*

The supplementary provisions were issued as part of amendment 3 in February 1988. They were proposed by the British Property Federation and there are marked similarities between the provisions and certain clauses in the BPF's own form of contract. The provisions are as follows:

S1 [Not used] This was adjudication, but following the Housing Grants, Construction and Regeneration Act 1996 Part II a more substantial clause which complied with the Act was required. Adjudication is now clause 39A.
S2 Submission of drawings, etc. to the employer.
S3 Site manager.
S4 Persons named as subcontractors in the Employer's Requirements.

S5 Bills of quantities.
S6 Valuation of change instructions - direct loss and/or expense - submission of estimates by the contractor.
S7 Direct loss and/or expense - submission of estimates by contractor.

The provisions will be dealt with throughout the book under the various topic headings.

The supplementary provisions are referred to in article 1 and, if the employer wishes them to apply, the appropriate part of appendix 1 must be completed. It is suggested that the employer would be wise to complete the appendix so that the provisions do apply, because they are generally very sensible. If some of the provisions are not required, they should be deleted. Contractors must be wary to see where they do apply and, if so, note every significant effect on the contract.

2.7 *Priority, discrepancies, errors*

Clause 2.2 provides that nothing in the Employer's Requirements, the Contractor's Proposals or the Contract Sum Analysis overrides or modifies the application or interpretation of anything which is in the articles of agreement, the conditions and where applicable, the supplementary provisions and the appendices. The effect in practice is that if there is a conflict between a term in the printed contract and a term in the Employer's Requirements, say differing periods of notice under clause 4.1.2, the period in the printed form will apply. This type of clause has been upheld in the courts: see, for example, *M.J. Gleeson (Contractors) Ltd* v. *Hillingdon Borough Council* (1970); *English Industrial Estates Corporation* v. *George Wimpey & Co Ltd* (1973). In the absence of this clause, the ordinary rule of interpretation would apply, namely that where a contract is contained in a printed form, and there is inconsistency between the printed terms and typewritten terms, the typewritten terms would prevail. That sensibly assumes that if the parties have a set of contract documents consisting of a standard printed form and a typed or written section, the typed or written section would prevail in the case of any conflict. In order to amend a printed clause it is necessary to amend it on the form itself and have both parties initial the amendment. Another way is to have any special clauses initialled by the parties and annexed to the printed form with an appropriate reference inserted in article 1.

The simplest way of removing the problem is to delete clause 2.2 altogether. Care must be taken not to fall into the trap of simply stating in the Employer's Requirements that the clause is deleted without actually deleting it in the form! It is very common and correct for any amendments to the contract clauses to be listed in the Employer's Requirements. Where this occurs, and if clause 2.2 is not deleted, the employer's professional advisor must ensure that these amendments are meticulously transferred to the printed form before the contract is executed.

Clauses 2.3 and 2.4 deal with discrepancies. Clause 7 provides that the employer must define the site boundaries. Clause 2.3.1 reasonably provides that if there is a divergence between what the employer has defined and anything contained in the Requirements, he must issue an instruction to correct the matter which is deemed to be a change and it is to be valued under clause 12. If either the employer or the contractor finds the divergence, he must give the other a written notice. It is now established that the contractor has no duty to look for or to find such divergences in this or other instances, but simply to give notice if he finds them: *London Borough of Merton* v. *Stanley Hugh Leach Ltd* (1985). If the contractor is himself to provide the site, clause 2.3.1 must be amended. It is suggested that the amended clause should make reference to the Contractor's Proposals instead of the Employer's Requirements and to the definition of site boundary given by the contractor.

Clause 2.4.1 deals with the position if there is a discrepancy within the Employer's Requirements or between the Requirements and any change issued in accordance with clause 12.2. The reference to the change is intended to cover the situation where the employer issues a change instruction which while obviously changing the particular part of the Requirements at which it is aimed, inadvertently conflicts with something else which is not the subject of the change. In the case of any such discrepancies, if the matter is addressed within the Contractor's Proposals, they will prevail and there will be no additional costs to the employer, neither will there be any reduction even if the treatment in the Contractor's Proposals is clearly less expensive than either of the discrepant items.

Complications can arise. For example, both walnut panelling and plastic faced steel panelling may be separately required for the boardroom. If the Contractor's Proposals allow for only plaster, a difficult situation can arise in which there is a discrepancy, not only within the Employer's Requirements, but also between them and the Contractor's Proposals. The solution to such a problem is to analyse it in terms of the discrepancy between documents (see below), then in terms of clause 2.4.1. If the Contractor's Proposals do not deal with the matter, the contractor is required to give the employer written notification of his amendment to resolve the discrepancy. The employer may either accept the amendment or decide on a different solution. In either case, the employer's decision is to be treated as a change which will be valued under clause 12.5. In addition, the contractor may be entitled to an extension of time under clause 25.4.5.1 and direct loss and/or expense under clause 26.2.6. If the decision by the employer is late and causes delay or disruption to the contractor, that again is grounds for both extension of time and loss and/or expense. It is thought that, if the discrepancy was not detected until the element was constructed, the employer must issue a change instruction to correct the problem and the contractor would be entitled to reimbursement in terms of money, time and loss and/or expense.

Clause 2.4.2 deals with discrepancies within the Contractor's Proposals. The contractor must immediately inform the employer in writing giving details of his suggested amendment. The employer may then choose between the discrepant items or he may choose the contractor's suggestion, all at no additional cost. The employer must take care to give a decision within a reasonable time or the contractor will have grounds for extension of time and loss and/or expense; this situation is expressly referred to in clauses 25.4.6 and 26.2.7 respectively. If the employer dislikes all the available options, he may issue a change instruction under clause 12.2, but that is not a prudent course unless absolutely necessary, because the employer pays a penalty in the cost of the change and, possibly, extension of time and loss and/or expense. The results of failure by either party to note the discrepancy before construction would be firmly at the cost of the contractor.

What is the situation if there is a discrepancy between the Employer's Requirements and the Contractor's Proposals? The contract is silent on this point. The third recital states that the employer has examined the Contractor's Proposals and the Contract Sum Analysis and, subject to the conditions and where applicable the supplementary conditions, he is satisfied that they appear to meet his requirements. This is not far from saying that the employer is absolutely certain that he is not quite sure. A footnote emphasises the importance of ensuring that the two documents are consistent. Practice Note CD/1B makes reference to the problem and it is worth quoting:

> 'The Contract Conditions do not deal with the position where, despite the Third recital and the advice in footnote [b], there is a divergence between the Contractor's Proposals and the Employer's Requirements. It was considered more appropriate to emphasise the need to follow the advice in footnote [b] than to include any specific provision on such divergences.'

In our view this equivalent of hoping for the best is not the most appropriate way to deal with the discrepancies between documents which will inevitably arise. One way to deal with the matter is simply to amend the third recital so that the employer is satisfied that the Contractor's Proposals do meet the Employer's Requirements. As it stands, the third recital is intended to indicate that the employer accepts that, at face value, the Contractor's Proposals respond to his stated criteria, but he is reserving his position as regards the actual satisfaction of such criteria. There is something to be said for this approach, because the employer, whether or not professionally advised, cannot be expected to carry out detailed checks of the Proposals. Any 'approvals' given by the employer must be seen in this light: *Hampshire County Council* v. *Stanley Hugh Leach Ltd* (1991). The simplest way to tackle the problem is to insert a clause to the effect that if there is any discrepancy between Employer's Requirements and the Contractor's Proposals, the Employer's Requirements will take precedence. This is probably the correct position even without an express term and the priority situation created by clause 8.1 reinforces that view (see section 4.2).

If the contractor makes a unilateral error in his Proposals or in the Contract Sum Analysis, e.g. errors in pricing, he will have to stand the consequences unless the employer or his professional advisors discover the error before acceptance and realise that it is not intentional: *W. Higgins Ltd* v. *Northampton Corporation* (1927). This may be thought a harsh view, but the contractor undertakes a very great burden of responsibility under this form of contract. After all, that is the attraction so far as the employer is concerned. The contractor may possibly get some relief if he can demonstrate that the employer knew of the error at the time the tender was accepted: *McMaster University* v. *Wilchar Construction Ltd* (1971), a decision of the Ontario High Court.

2.8 *Custody and copies*

Clause 5.2 provides that immediately after the execution of the contract, the employer must provide the contractor with one copy of each of the articles of agreement, the conditions, the appendices, the Employer's Requirements, the Contractor's Proposals and the Contract Sum Analysis. Each document must be certified on behalf of the employer (therefore, presumably by his agent). It is quite sufficient for the employer's agent to write on each document: 'Certified a true copy of the...' and sign and date it. Custody of the Employer's Requirements and the Contractor's Proposals is to be the responsibility of the employer in accordance with clause 5.1 although, curiously, there is no mention of articles of agreement. Reading clauses 5.1 and 5.2 together leads to the conclusion that all original documents are to be in the custody of the employer although it is surprising that the clause does not specifically state this. In practice it should not be of any great importance as long as the contractor is provided with a certified copy at an early date. Sometimes the employer may fail to provide the certified copies and employers have been known to refuse to provide them even when requested. The prudent contractor will always keep a copy of the documents he signs against just such an eventuality. The employer's failure would no doubt be susceptible to a mandatory injunction if the contractor had no copy and the employer persistently refused to supply one, but in the event of a dispute which went to arbitration or litigation, the documents would be subject to discovery.

The contractor has a duty under clause 5.3 to supply the employer with two copies of all the drawings, specifications and other information which he either prepares or uses for the purposes of the Works. Thus, strictly, the contractor must supply information prepared for the Works even if ultimately unused. Despite what may be thought, this information is not supplied for the purposes of securing the employer's approval. It is purely for information or record purposes. If the employer quite reasonably desires the opportunity of commenting on the information before it is used, supplementary provision S2 must be used and the employer should include requirements for submission, comments and timing in the Employer's Requirements (see section 4.4.1). The contractor must keep one copy of all this information, together with the Employer's Requirements, the Contractor's Proposals and the Contract Sum Analysis on site so that the employer's agent can have access to them at all reasonable times (clause 5.4).

Clause 5.5 is most important because it is referred to in the determination provisions and it is vital to understand exactly what it means. It stipulates that 'before commencement of the Defects Liability Period', i.e. before practical completion, the contractor must supply the employer with whatever drawings and information are specified in the Employer's Requirements and the Contractor's Proposals relating to the Works as built and their maintenance and operation including any installations (presumably such things as heating systems). The important point is that the contractor's obligation is essentially to supply what are commonly known, and indeed noted in the margin, as 'as-built' drawings (see Chapter 12). The usual prohibition against divulging the contents of contract documents and other confidential information is put on both parties by clause 5.6 with the exception of any information which the employer wishes to use in connection with maintenance, use, repair, advertisement, letting or sale of the Works.

2.9 *Notices*

Clause 1.5 of the contract sets out the requirements for the giving or service of notices or other documents. It applies if the contract does not expressly state the way in which service of documents is to be achieved. Therefore, it does not apply to notices given in connection with the determination procedures in clauses 27, 28 and 28A, because those clauses state that service is to be carried out by means of actual delivery, special or recorded delivery. In other cases, service is to be by any effective means, to any agreed address. Surprisingly, parties are quite capable of squabbling over service and appropriate addresses. If that is the situation, service can be achieved by addressing the document to the last known principal business address or if the addressee is a body corporate, to that body's registered office or its principal office provided it is prepaid and sent by post.

The contract also provides for electronic data interchange if the parties so wish. If they do so wish, they can insert an appropriate reference into appendix 1 and then clause 1.8 states that supplemental conditions for EDI apply. These conditions are bound in towards the back of the contract. This is a very sensible option and it is in line with current practice in many (although by no means all) offices.

In broad terms, the supplementary conditions provide that the parties will enter into an EDI agreement no later than the date on which a binding contract comes into existence between the employer and the contractor. In practice, the parties will execute the building contract and the EDI agreement at the same time. Clause 2 states that dispute resolution procedures under the building contract are to apply to the EDI agreement and they will prevail over any dispute resolution procedures in the EDI agreement.

The EDI agreement cannot override or modify anything in the contract unless the provisions expressly so state. Provision 1.2 is not very clear. It states that the types and classes of communication to which the EDI agreement applies, and persons between whom exchanges are to be made, are as stated in the contract documents or subsequently agreed between the parties in writing. The appendix says nothing other than whether the agreement applies and the type of agreement to be used. The provisions permit communications which the contract requires to be in writing to be exchanged electronically. There are specific exceptions which must always be in writing. They are:

- Determination of the contractor's employment
- Suspension by the contractor of his obligations
- The final account and final statement and the employer's final account and employer's final statement
- Invoking dispute resolution procedures, for example, a notice to concur in the appointment of an arbitrator
- Any agreement which the parties may enter into which amends the contract, including the EDI provisions.

Clause 1.6 usefully sets out the way in which periods of days are to be reckoned. This is to comply with the Housing Grants, Construction and Regeneration Act 1996. If something must be done within a certain number of days from a particular date, the period begins on the day after that date. Days which are public holidays are excluded. Clause 1.3 helpfully defines public holidays as 'Christmas Day, Good Friday or a day which under the Banking and Financial Dealings Act 1971 is a bank holiday'. A footnote instructs the user to amend the definition if different public holidays apply.

Clause 1.7 states that the law applicable to the contract is to be the law of England no matter that the nationality, residence or domicile of any of the parties is elsewhere. Where a different system of law is required, this clause must be amended. For example, if the Works are being carried out in Northern Ireland, the parties will probably wish the applicable law to be the law of Northern Ireland. Curiously, the applicable law of the two bonds which are now bound into the contract is stated to be the law of England and Wales.

CHAPTER THREE
DESIGN LIABILITY

3.1 *General principles of design liability*

3.1.1 Basic principles of liability

The liabilities of a designer are in principle no different from the liabilities of any person, i.e. the designer may have liabilities in contract or in tort.

The general principle of contract is that the parties to the contract have agreed on mutual rights and duties which they would not otherwise have in law. The parties may agree any terms they wish provided only that they are lawful. The terms of a contract between an architect and his client will usually include an express term that the architect will use reasonable skill and care in the execution of his duties under the contract. Such a term will usually be express, but it may also be implied since anyone holding himself out as an architect impliedly warrants that he possesses the necessary ability and skill. Terms may be implied into a construction contract or indeed into a contract for professional services either as a matter of law or as a matter of fact (see section 4.2).

The classic modern statement on the doctrine of implication of terms is that of Lord Simon in *BP Refinery Ltd* v. *Shire of Hastings* (1978), where he said :

> '[For] a term to be implied, the following conditions (which may overlap) must be satisfied: (1) It must be reasonable and equitable; (2) it must be necessary to give business efficacy to the contract, so no terms will be implied if the contract is effective without it; (3) it must be so obvious that "it goes without saying"; (4) it must be capable of clear expression; (5) it must not contradict any express term of the contract.'

An architect's duties may be quite extensive and include far more than just design (see, for example, the RIBA Standard Form of Agreement for the Appointment of an Architect (SFA/99)). In the case of design and build, typical design contracts would be between an architect and his client in the early stages of a project, then in the later stages of a project, between the contractor and the employer and possibly between the contractor and an architect or other construction professional. Any formal collateral warranty entered into is also a contract (see section 3.6).

A failure by a party to a contract to carry out his duties regarding design would be a breach of contract which would entitle the other party to recover damages in respect of any proven loss, and it matters not whether the designer is an architect or other design professional or a design and build contractor. The principle of recovery is that the injured party should be put in the position, so far as money can, as if the breach had not been committed (*Robinson* v. *Harman* (1848)) although this may be modified in practice: *Ruxley Electronics & Construction Ltd* v. *Forsyth* (1995). In contract, there are two kinds of damage which can be recovered:

- Damages that may fairly and reasonably be considered to arise naturally out of the breach; *and*
- Damages which are the result of special circumstances known to the parties at the time the contract was entered into and which are capable of causing a greater loss than otherwise would be the case: *Hadley* v. *Baxendale* (1854).

A designer whose design fails may also be liable in tort for negligence. In light of recent developments in the law, it is thought that where there is a contract between the parties, there will be a parallel liability owed in tort to the other party for any resulting loss which may be wider in scope than the liability in the contract: *Holt* v. *Payne Skillington* (1995). This is so whether the negligent designer is an architect or other design professional or a design and build contractor.

Economic loss is distinct from damage which results from physical injury to or death of a person or physical damage to property other than the building itself.

Until quite recently, it was common for a building owner whose building failed to bring actions in both contract and in tort, the one claiming damages for breach of contract, and the other claiming damages for negligence. It is sometimes difficult to decide whether a particular loss is to be categorised as economic and the courts have always tended to make the decision on the grounds of policy as much as anything else, because in the last analysis, it is possible to say that most loss is economic: *Spartan Steel & Alloys Ltd* v. *Martin & Co (Contractors) Ltd* (1972).

The sea change came about in *Murphy* v. *Brentwood District Council* (1990) and, as a result of that case, it has been well said that it is now 'necessary to read all English authorities concerning negligence decided between 1971 and 1990 with extreme caution': *Keating on Building Contracts*, 6th edn, 1995, p.169.

However, although it is now clear that normally a plaintiff cannot recover economic loss in an action for negligence, but must establish actual death or physical injury to persons or property other than the defective building itself, economic loss can be recovered 'where there is a special relationship amounting to reliance by the plaintiff on the defendant or where the economic loss is truly consequential upon actual physical injury to person or property': *Keating*, op.cit., p.170.

A 'reliance situation' will seemingly only arise if there is a special relationship of both proximity and reliance between the parties. Where the representor has some special skill or knowledge about which he gives advice, and he knows, or it is reasonably foreseeable, that the other will rely on the advice and the advice has been acted on by the other, the resultant economic loss will be recoverable provided that it is foreseeable: *Hedley Byrne & Co* v. *Heller & Partners Ltd* (1964).

The House of Lords have reconsidered and restricted the criteria for the special relationship. The current position appears to be that a special relationship will be considered to exist if:

- The advice is required for a purpose which is made known to the misrepresentor at the time the advice is given; *and*
- The misrepresentor knows or can reasonably foresee that the advice will be communicated to the other either personally or as a member of an ascertainable class in order to be used for the purpose initially made known; *and*
- It is known by the misrepresentor that the advice is likely to be acted on by the other without further enquiry; *and*
- It is so acted on and the other suffers some detriment: *Caparo Industries plc* v. *Dickman & Others* (1990).

Because special skills and advice are features of this kind of liability, it is often associated with professional advice, although not necessarily: *Barclays Bank* v. *Fairclough Building Ltd and Carne (Structural Repairs) Co Ltd and Trendleway Ltd* (1995).

At one time, the case of *Junior Books Ltd* v. *The Veitchi Co Ltd* (1982) appeared to open the floodgates to the recovery of economic loss, but in subsequent cases it has been distinguished almost to the point of extinction: *Muirhead* v. *Industrial Tank Specialities Ltd* (1986), although it has not been directly overruled and, indeed, in *Murphy* v. *Brentwood District Council* (1990) it was explained by some members of the House of Lords as resting on the *Hedley Byrne* principle of reliance, although on what basis that 'reliance' arose is not easy to see. The House of Lords appeared to treat the employer/nominated subcontractor relationship as an almost unique situation.

There was no collateral contract involved in that case, which involved a nominated subcontractor - and the relationship between a nominated subcontractor and an employer is hardly, on any reasonable view, a 'special relationship'. It is thought that the decision is not of general application and would certainly not be extended.

The current position as determined by the House of Lords in the important case of *Murphy* v. *Brentwood District Council* (1990) may be broadly summarised, in the context of the construction industry, as follows:

- Negligence which results in a defect in the building itself is not actionable in tort. There is no actionable damage to 'other property'.
- To be actionable, the defective structure must cause damage to other property or result in death or personal injury.
- If, however, the defect is discovered before it has caused damage, the cost of making good the defect is not recoverable.
- In a complex structure such as a building, it is not permitted to consider the building as a series of segments, one causing damage to another - the so-called 'complex structure' theory.
- However, there is possibly potential liability where 'some distinct item incorporated in the structure... positively malfunctions so as to inflict positive damage on the structure in which it is incorporated'. For example, where a subcontractor instals a central heating boiler into a building and it explodes causing damage to the building, the negligent installer might be held liable in damages or where subsequent underpinning is the cause of damage to the original building: *Jacobs* v. *Morton & Partners* (1994).

The reasoning which led up to the *Murphy* decision was given earlier effect in *Pacific Associates* v. *Baxter* (1988), where contractors were unsuccessful in an action for negligence against a supervising engineer under a FIDIC contract who had failed to certify certain sums allegedly due to them. The contractors also alleged unsuccessfully a breach by the engineers of a duty to act impartially.

The Court of Appeal, in dismissing the contractors' claim, stressed the importance of the terms of the contract between the employer and the contractor which provided for arbitration of disputes arising under the contract.

In stressing the importance of the contractual route, the following principles appear to have been important:

- The engineers were agents of the employer.
- The engineers had a contractual obligation to the employer to use skill and care and to act fairly between the parties.
- The contractors had relied on their remedies against the employer under the contract between them by going to arbitration over the disputed claims. In the event, the arbitration was settled on terms of an *ex gratia* payment by the employer to the contractors who were attempting to recover the shortfall between the amount claimed and the amount of the settlement by means of an action in tort against the engineers.
- The engineers had not assumed responsibility to the contractors for economic loss resulting from a breach of any of the obligations in the contract between the employer and the contractors.
- Therefore, there was no basis, either on the *Hedley Byrne* principle or otherwise, by which the engineers could be said to owe duty of care to the contractors.
- There was an arbitration provision which enabled the contractor to recover from the employer. The engineers had a duty to act in accordance with the construction contract, but that duty arose from the contract between the engineers and the employer. The contractors could challenge the performance of the engineers by claiming against the employer. There was, therefore, no justification for imposing on the contractual structure an additional liability in tort.

This is of great importance to architects and others advising either employer or contractor and the principles were also applied by the High Court of Hong Kong to an architect under a building contract.

However, it should be noted that *Pacific Associates* puts weight on the existence of a contractual route for the contractor to obtain redress, i.e. arbitration, and it is still possible that, where the arbitration clause is missing or the employer is in liquidation, the contractor might be able to recover against a negligent or unfair or partial architect or other certifier who deliberately or negligently undercertifies amounts due.

Recent developments in the way the courts have interpreted the

Hedley Byrne principle suggest that actions against architects based on reliance may not be far away. It has been held that the principle extends beyond the provision of information to include the performance of other services: *Henderson* v. *Merrett Syndicates Ltd* (1994); and that there is no sustainable distinction between the making of statements and other exercises of duty: *Conway* v. *Crowe Kelsey & Partner* (1994).

3.1.2 Designer's position

The professional designer, such as an architect or an engineer, is required to exercise reasonable skill and care: *Lanphier* v. *Phipos* (1838). He is not required to guarantee the result unless he also provides the end product, such as a dentist making a set of false teeth: *Samuels* v. *Davis* (1943). A designer who fails to exercise the requisite amount of skill and care may be negligent:

> 'Where you get a situation which involves the use of some special skill and competence, then the test as to whether there has been negligence or not is not the test of the man at the top of the Clapham omnibus, because he has not got this special skill. The test is the standard of the ordinary skilled man exercising and professing to have that skill; it is well established law that it is sufficient if he exercises the ordinary skill of an ordinary competent man exercising that particular art.' (*Bolam* v. *Friern Hospital Management Committee* (1957))

Thus, a designer will be thought to have acted correctly if he acts with the kind of skill an average designer would display. In order to decide such issues, the court must hear expert testimony from other designers on the matter in question. It is not always sufficient for a designer to be able to maintain that he simply did the same as other designers were doing, if it can be shown that generally accepted practice is not correct: *Sidaway* v. *Governors of the Bethlem Royal Hospital and the Maudsley Hospital* (1985).

A person will be judged on the basis of the skills he professes himself to have. Therefore, if an architect holds himself out to have special expertise in the restoration of old buildings, his failure in that respect will be compared to the performance of other architects who have that special expertise, whether or not the original architect does in fact have such expertise.

On the other hand, the standards to be applied to a professional person will be the standard of the time the professional acted and not the standards commonly practised at some time after the act: *Wimpey Construction UK Ltd* v. *D.V. Poole* (1984). This is usually referred to as the 'state of the art' defence.

Moreover, the designer's responsibility does not end when the design is completed and handed to the builder. He has a continuing responsibility to review and revise the design if he becomes aware of any problems: *London Borough of Merton* v. *Lowe & Pickford* (1981). It appears that this duty ends after the designer's initial involvement ends so that the designer is not burdened with the responsibility of constantly reviewing designs thereafter: *T.E. Eckersley and Others* v. *Binnie & Partners and Others* (1988).

The designer who uses untried methods of construction or materials will be just as liable for design failures as if he fails while using well tried methods and materials. He cannot blame general lack of knowledge as a basis for a state of the art defence. Special care is needed, therefore, before new techniques are put into operation:

> 'For architects to use untried or relatively untried materials or techniques cannot in itself be wrong, as otherwise the construction industry can never make any progress. I think however, that architects who are venturing into the untried or little tried would be wise to warn their clients specifically and get their express approval.' (*Victoria University of Manchester* v. *Hugh Wilson & Lewis Womersley and Pochin (Contractors) Ltd* (1984))

3.1.3 Fitness for purpose

As already mentioned, the law will require a professional person, such as an architect or an engineer, to exercise reasonable skill and care in the performance of his duties. This standard is also required of the designer by the Supply of Goods and Services Act 1982 in respect of any contract for the supply of design services. This statutory duty can be displaced by the imposition of a stricter duty in a contract.

The stricter duty is normally what is known as 'fitness for purpose'. Such a duty may be expressly stated in a contract in those words, or words to the same effect, or the law will usually imply it where the contract is on the basis of work and materials unless it is clear that the employer is not relying on the contractor: *Young & Marten* v. *McManus Childs Ltd* (1968).

Where an employer relies on a contractor to provide an entire building and there is no independent designer involved, a term of reasonable fitness for purpose will be implied irrespective of any negligence or fault or whether the unfitness results from the quality of work or materials or from defects in the design: *Viking Grain Storage Ltd* v. *T.H. White Installations Ltd* (1985). The suggestion that matters of design in such circumstances should be regarded as involving no more than reasonable care was rejected by the House of Lords in the television mast case, *Independent Broadcasting Authority* v. *EMI Electronics Ltd and BICC Construction Ltd* (1980):

> 'As they undertook responsibility for the design, they became contractually responsible to ITA for BICC's negligence, and so in my opinion are liable to ITA in damages for breach of contract. In the circumstances it was not necessary to consider whether EMI had by their contract undertaken to supply a mast reasonably fit for the purpose for which they knew it was intended and whether BICC had by their Contract with EMI undertaken a similar obligation but had that been argued, I would myself have been surprised if it had been concluded that they had not done so.'

A fit for purpose design liability might sometimes be implied into a designer's contract. *Greaves & Co Contractors* v. *Baynham Meikle & Partners* (1975) is a case in point which dealt with the liability of an engineer carrying out the design of a warehouse floor for a contractor who was engaged by the employer on a design and build basis. Lord Denning said:

> 'The law does not usually imply a warranty that he will achieve the desired result but only a term that he will use reasonable skill and care. The surgeon does not warrant that he will cure the patient. Nor does the solicitor warrant that he will win the case. But, when a dentist agrees to make a set of false teeth for a patient, there is an implied warranty that they will fit his gums, see *Samuels* v. *Davis* (1943).
>
> What then is the position when an architect or an engineer is employed to design a house or a bridge? Is he under an implied warranty that, if the work is carried out to his design, it will be reasonably fit for its purpose or is he only under a duty to use reasonable skill and care? This question may require to be answered some day as a matter of law. But, in the present case I do not think we need answer it. For the evidence shows that both parties were of one mind on the matter. Their common intention was that the engineer should design a warehouse which would be

> fit for the purpose for which it was required. That common intention gives rise to a term implied in fact.'

In that case, fitness for purpose was not implied as a matter of law, but as a matter of fact, i.e. both parties had intended it to be so. In a situation where a JCT traditional standard form is being used, the contractor will have no design liability: *John Mowlem & Co Ltd* v. *British Insulated Callenders Pension Trust Ltd* (1977):

> 'I should require the clearest possible contractual condition before I should feel driven to find a contractor liable for a fault in the design, design being a matter which a [designer] is alone qualified to carry out.'

Such a condition is, of course, available in a limited way for performance specified work in clause 42 of JCT 98.

3.1.4 Contractor's responsibility to warn

The question often arises as to the contractor's duty to warn the employer or the architect if he finds defects in the design. Considering the JCT Standard Form 1963 Edition, which requires the contractor to notify the architect if he finds any discrepancy in or between the documents, it was held that although the contractor has a duty under that form to notify discrepancies, he has no duty to look for and find the discrepancies in the first instance: *London Borough of Merton* v. *Stanley Hugh Leach Ltd* (1985). In the Canadian case of *Brunswick Construction Ltd* v. *Nowlan* (1974), a contractor charged with carrying out the construction of a house to architect's designs was held to be liable for an error in the design where the original architect was not engaged to inspect the work. In this, as in other cases, the question of reliance appears to be important.

In *Equitable Debenture Assets Corporation Ltd* v. *William Moss* (1984), it was held that to give efficacy to the contract, a term was to be implied requiring the contractor to warn the architect of design defects and that there was a duty of care in negligence to the employer and to the architect in this regard. The point was emphasised in *Victoria University of Manchester* v. *Hugh Wilson & Lewis Womersley and Pochin (Contractors) Ltd* (1984):

> 'In this case, I think that a term was to be implied in each contract requiring the contractors to warn the architects as the Univer-

> sity's agents of the defects in design, which they believed to exist. Belief that there were defects required more than mere doubt as to the correctness of the design, but less than actual knowledge of errors.'

The point was given further consideration in *University of Glasgow* v. *W. Whitfield and John Laing (Construction) Ltd* (1988) where it was held that the decisions in *EDAC* and *Manchester* were concerned with the duty of a contractor to warn the employer, not with any duty owed by the contractor to warn the architect. Both cases assumed a special relationship of reliance between the contractor and the employer. Some doubt may have been thrown on this decision, however, when Judge Newey QC, who was responsible for the decisions in *EDAC* and *Manchester*, returned to the theme in *Edward Lindenberg* v. *Joe Canning & Jerome Contracting Ltd* (1992):

> 'In *Brunswick Construction* the Supreme Court of Canada held that experienced contractors had acted in breach of contract in building a house in accordance with plans prepared by an engineer in which they should have detected defects. In *Equitable Debenture* I held that there was an implied term in a contract requiring contractors to inform their employer's architect of defects of design of which they knew and in *Victoria Manchester University* I held that the implied term extended to defects which they believed to exist.'

This is a point of some importance in the light of any gaps perceived to exist between the design obligation owed by the original architect and the contractor under WCD 98. In *Bowmer & Kirkland* v. *Wilson Bowden Properties Ltd* (1996), the court observed that it was 'a feature of good workmanship for a contractor to point out obvious errors, or if there is doubt or uncertainty in the plans, specification or other instructions, to ask for clarification so that the uncertainty is removed'.

3.2 *Liability under the contract*

The most important difference between WCD 98 and JCT 98 is the contractor's obligation to design as well as construct. Virtually all the other differences spring from this central obligation. The design obligation deserves careful scrutiny. It is to be found in article 1 and clauses 2.1 and 2.5.

Article 1

The contractor is to 'both complete the design for the Works and carry out and complete the construction of the Works'. The Works are defined as 'the works briefly described in the First Recital and referred to in the Employer's Requirements and the Contractor's Proposals and including any changes made to those works in accordance with this Contract'. This duty is 'subject to the Conditions and, where so stated in Appendix 1,' to the Supplementary Provisions. It is to be noted that the contractor is not to design, but to complete the design.

Clause 2.1 Contractor's obligations

This clause is so important that the first part must be quoted in its entirety:

> 'The Contractor shall upon and subject to the Conditions carry out and complete the Works referred to in the Employer's Requirements, the Contractor's Proposals (to which the Contract Sum Analysis is annexed), the Articles of Agreement, these Conditions and the Appendices in accordance with the aforementioned documents and for that purpose shall complete the design for the Works including the selection of any specifications for any kinds and standards of the materials and goods and workmanship to be used in the construction of the Works so far as not described or stated in the Employer's Requirements or Contractor's Proposals.'

The contractor's obligation is split into two distinct parts. First he must carry out and complete the Works in accordance with the contract documents. This would be the contractor's normal obligation under traditional forms such as JCT 98 or IFC 98. His second obligation follows. In the first edition, we commented that its method of expression was very strange. The contractor is to complete the design of the Works, but the Works are restricted to those referred to in the Employer's Requirements and the Contractor's Proposals. The contractor seems to have no liability under this clause to design changes. Article 1 does not assist in widening design responsibility because it is made 'subject to the Conditions', of which clause 2.1 is the principal.

Moreover, the contractor is to complete the design for the express purpose of carrying out and completing the Works in accordance with the contract documents and not such documents modified by the effect of any change instruction. JCT Amendment 7 added a sentence to the clause which assists in setting out, but does not fully clarify, that changes are included. The contractor is not given responsibility for the design as a whole, but merely to complete what is, presumably, left incomplete. It therefore follows that if the employer, for whatever reason, has caused the whole of the design to be prepared by an independent architect and included within the Employer's Requirements, there will be no design to complete and the contractor's obligation in this regard will be non-existent. At the other end of the same spectrum, if there is no design included in the Employer's Requirements, the contractor will be responsible for the whole of the design.

The scheme of the contractor's obligations is clear. He is to construct the whole of the Works, but only to design whatever is left undesigned by the Employer's Requirements. Because most contracts let on this particular form include some design input on behalf of the employer, this clause immediately introduces the very thing which the philosophy of design build is intended to avoid: uncertainty regarding design responsibility.

The second part of clause 2.1 is ambiguous. It has at least two possible meanings:

- The contractor must complete the design of the Works only to the extent that it is not described or stated in either the Employer's Requirements or the Contractor's Proposals and the design is to include the selection of any specifications for any kinds and standards of the materials and goods and workmanship to be used in the construction of the Works; *or*
- The contractor must complete the design of the Works and must include the selection of any kinds and standards of the materials and goods and workmanship to be used in the construction of the Works if they are not described or stated in the Employer's Requirements or the Contractor's Proposals.

It is not absolutely clear whether it is the design, or the selection of any specifications which may be in the Employer's Requirements or the Contractor's Proposals. Commonsense suggests that it is the first meaning which was intended by the JCT in drafting this clause. However, at best it seems to state the obvious and, at worst, to throw doubt on the precise extent of the contractor's obligations under this clause.

Clause 2.5.1 Contractor's design warranty

This clause actually sets out the contractor's design liability. From the contractor's point of view, it is a most important clause and a good reason why a contractor should always opt for this contract rather than a simple exchange of letters, under which he would be under a 'fitness for purpose' liability unless the lower standard was specified.

Under Clause 2.5.1. the contractor is to have the same liability to the employer as an architect or appropriate professional designer holding himself out as competent to do the design. The clause goes further and, lest there be any doubt, makes clear that it is referring to an architect or designer acting independently under a separate contract with the employer having supplied the design for a building to be carried out by a contractor who is not supplying the design. The liability to which this clause refers is the liability of an averagely competent professional who is liable only to the extent that he fails to exercise reasonable skill and care (see section 3.1.2 above) and it is a mystery why it did not say so instead of using the present clumsy form of words.

This should be contrasted with the liability of a contractor that the building which he has designed and built is reasonably fit for any purpose which has been made known to him (see section 3.1.3 above). The contractor's liability under this form of contract is considerably less than the liability he would shoulder at common law. To that extent this contract represents a very valuable restriction of the contractor's design liability. In view of the negotiated status of this contract, this restriction is not one which falls under the provisions of the Unfair Contract Terms Act 1977.

The liability is expressed to be the same as a professional 'whether under statute or otherwise'. In other words, the liability to the employer is the same as any professional person under the Supply of Goods and Services Act 1982, section 13 - reasonable care and skill - and is to be the same professional standard of care in tort also. It is unlikely that a court would find a higher standard of care applicable in tort where the parties had the opportunity to express such standard in their contract and they had not taken it (see section 3.1.1 above). The contractor's liability is in respect of any defect or insufficiency in the design.

The clause very precisely sets out the boundaries of the design to which this professional design liability extends and they may be listed as:

- The design comprised in the Contractor's Proposals
- The design which the contractor is to complete under clause 2
- The design which is in accordance with the Employer's Requirements
- The design which is in accordance with the conditions including any further design which the contractor is to carry out as a result of a change instruction.

At first sight, this provision covers all the design work which the contractor may do other than what has been carried out by or on behalf of the employer and incorporated in his Requirements. It seems, therefore, that if there is any design which the contractor is to carry out under the contract, but which can be brought outside the boundaries listed above, the contractor will have the ordinary liability of fitness for purpose in respect of such design.

Before the sentence was added to clause 2.1 by Amendment 7, the design involved in a change instruction was clearly excluded, as we noted in the first edition, and it was only further design to be carried out as a result of such instruction which was included. This corresponded with the contractor's liability for design under clause 2.1 as discussed above.

Clause 2.5.2

This clause makes clear that if the contract involves any work in connection with dwellings, the contractor has liability under the Defective Premises Act 1972. The key provision of the Act is to be found in section 1(1) which provides:

> 'Any person taking on work for or in connection with the provision of a dwelling ... owes a duty to see that the work which he takes on is done in a workmanlike or, as the case may be, professional manner, with proper materials and so ... that the dwelling will be fit for human habitation when completed.'

The obligation put on professionals should be noted. The contractor's design obligation under clause 2.5.1 is said to be that of an independent professional designer and it, therefore, amounts to the duty to use reasonable skill and care except when dwellings are involved, when the Act bites and the obligation becomes, to all intents and purposes, to design so as to be fit for purpose.

The clause provides that if the Employer's Requirements state that section 2(1) of the Act is to apply, the contractor and the employer must do everything necessary for the appropriate documents to be issued for the purposes of the section and the scheme approved under the Act. This section provides that section 1 of the Act does not apply to dwellings covered by an approved scheme.

Before 31 March 1979, dwellings sold with the National House Building Council (NHBC) guarantees were excluded from the Act, since the NHBC scheme was then approved. The NHBC scheme ceased to be approved after 31 March 1979 and consequently the Act applies to all persons taking on work for or in connection with the provision of a dwelling. As will be noted, the standard required by section 1(1) equates broadly to that required by the corresponding duties at common law.

The 1972 Act did not have retrospective effect *(Alexander* v. *Mercouris* (1979)). In *Thompson* v. *Clive Alexander & Partners* (1992) it was emphasised that the reference to fitness for habitation in section 1(1) merely sets the standard required in the performance of the duty created by section 1(1) and is not aimed at trivial defects.

Section 1 was considered by the Court of Appeal in *Andrews* v. *Schooling* (1991). In that case the plaintiff, the long lessee of a flat, claimed damages for breach of duty under section 1 on the grounds that the converted flat was not fit for human habitation. The flat suffered from damp caused by the evaporation of moisture from the cellar. The Court of Appeal held that section 1 applied to both acts of commission ('misfeasance') and omission ('non-feasance') and that it was irrelevant that the problem did not become apparent until after completion. In our view, the 1972 Act will become increasingly important in light of the restrictive decision of the House of Lords in *Murphy* v. *Brentwood District Council* (1990), especially as the NHBC scheme is no longer approved and there is no other approved scheme.

Clause 2.5.3 Limit of the contractor's liability

This clause gives opportunity for the employer to impose a limit on the contractor's liability in certain circumstances. An overriding proviso is that limitation on liability can apply only so far as it does not concern the contractor taking on work in connection with dwellings. That is to say that if the contract is for the design and construction of twelve flats, there can be no limitation of liability. If, however, the contract is for the design and construction of forty houses and a school, liability could be limited in respect of the school.

In so far as dwellings are not involved, the employer may

stipulate the limit of the contractor's liability for loss of use, loss of profit or other consequential loss arising in respect of the liability referred to in clause 2.5.1, i.e. design liability. This is a protection for the contractor and, in appropriate cases, the employer may state a limit in the hope that the reduced risk will be reflected in the tender prices. Alternatively, such a limit may be the result of negotiation following a high initial tender or as part of the usual two stage tendering procedure. The wording of the clause makes clear that the limit is to be stated in appendix 1 and that if no amount is therein stated, there will be no limit on liability under these heads. It seems, therefore, that if the Employer states in his Requirements that there will be limit on the contractor's liability under this clause of £25,000, the contractor's liability will still be unlimited if the employer neglects to insert the amount in appendix 1. This is because of the operation of clause 2.2 giving precedence to the articles, the conditions and the appendix.

The effect of this provision is probably rather less in practice than might appear to be the case at first sight. It really protects the contractor only 'from claims for special damages which would be recoverable only on proof of special circumstances and for damages contributed to by some supervening cause': *Saint Line* v. *Richardsons, Westgarth & Co Ltd* (1940). If a limit is stipulated in the appendix, it has no effect on the employer's power to deduct liquidated damages under clause 24.2.1 for failure to complete the Works by the completion date, and the deduction of such damages has no effect on the limit.

Clause 2.5.4

This clause simply makes clear that references to the design which the contractor prepares or issues include design prepared or issued by other persons such as independent firms of consultant architects specially commissioned by the contractor for the purpose. This is permitted by clause 18.2.3 (see Chapter 6).

3.3 *Design liability optional arrangements and consequences*

The architect's role was briefly discussed in section 1.3. The architect and other members of the design team may be employed either by the employer or by the contractor. Although they may be either in-house or independent, for simplicity's sake it is assumed here that the consultants are all independent. The position of the in-house designer, where different, is discussed further in section 3.5 below. Besides or instead of the architect, there may be several other consultants employed by the contractor. It is helpful to examine liabilities in three parts:

- Where the consultant carries out the whole of the design, or completes the design, for the contractor;
- Where the consultant produces the first part of the design, including the Employer's Requirements, for the employer;
- Where the consultant acts for the employer to obtain tenders, advises on the best tender to accept, puts together the contract documents and acts as employer's agent until completion of the project.

In the first instance while the consultant is acting for the contractor, he will owe such duties to the contractor as are expressly noted in whatever contractual agreement has been executed between them. There are now standard conditions of engagement published by the RIBA (SFA/99 and CE/99, each with appropriate amendment) suitable for use by architects carrying out work for a contractor in a design and build situation. If it is intended to use the standard forms SFA/99 or CE/99, care must be taken to use the appropriate amendments suitable for a design and build situation.

It is probable that the consultants also owe the contractor a duty of care in tort. Where WCD 98 is used, each consultant will be a subcontractor to the contractor under the provisions of clause 18.2.3. The contractor's liability to the employer will be the obligation to carry out the design using reasonable skill and care and the consultant should have a similar liability under his agreement with the contractor. Where the contractor employs a consultant who employs one or more subconsultants, each subconsultant is likely to have a duty in tort to the contractor in accordance with *Hedley Byrne* principles: *Cliffe Holdings* v. *Parkman Buck Ltd and Wildrington* (1996).

If a different kind of design and build contract is used (such as ACA 2), the contractor will probably have the higher liability of fitness for purpose and the contractor must ensure that the consultant has a similar level of liability. The problem here is that the consultant's professional indemnity insurers are virtually certain to reject any suggestion that the consultant takes on this higher level of liability and, without insurance, the higher liability has little practical value if a large claim is involved. Contractors must be wary of this point. In practice, the consultant's ordinary standard of liability is normally all that is required provided he is made aware of all relevant criteria before he commences the design. As in the case of *Greaves & Co (Contractors) Ltd* v. *Baynham Meikle & Partners* (1975), it is difficult for an engineer to contend that he should not be liable for a floor which is not fit for the purpose made known before the design was carried out. If, on receipt of the information, the engineer designed the floor using the same standard of reasonable skill and care to be expected from an ordinary competent engineer used to doing similar work, it should be fit for purpose.

Preparing design and constructional drawings for a contractor may be rather confusing at first for a consultant who has no experience of the design and build method of procurement. As noted above, his precise duties will depend principally on the terms of engagement. In general, however, some principles can be stated. The consultant is only empowered to do what is expressly agreed in his conditions of engagement. This may seem a very basic point, but to a consultant used to working in a traditional situation, it may come as a shock. For example, the contractor's design obligation is to complete the design and in doing so satisfy the Employer's Requirements. How far this obligation is transferred to the consultant is a matter for the contractor. If the Requirements ask for a heating system capable of producing a certain level of temperature under certain conditions, the contractor nevertheless may require the consultant to design a system giving more or less than this temperature or the contractor may set a totally different set of criteria to be satisfied. The consultant's duty is to satisfy the criteria communicated to him by the contractor who may have his own reasons for amending parts of the Employer's Requirements.

If the consultant is the architect charged with overall design work on behalf of the contractor, similar considerations apply. If the architect becomes aware that the employer has requested certain changes under the provisions of clause 12, he has no authority to incorporate the results of such changes in his design unless the contractor expressly passes on instructions to that effect. That is not to say that if a consultant becomes aware of such matters he should not inform the contractor; he should at least do that to protect his own position. The contractor may simply have made a mistake. But once notified, the consultant's duty to the contractor in respect of that particular matter is complete unless and until the contractor gives further instructions. Difficult questions may arise if the contractor asks the consultant to include in his production information matters which are contrary to the Building Regulations or to good practice. Although the consultant must take the contractor's instructions, he may not do anything contrary to law and there may be other circumstances where the consultant may deem it better to terminate his involvement under the conditions of appointment. Professionals cannot just blindly obey instructions. They have a broad and obvious duty not to act against their own professional judgment. Any professional in doubt about that should simply ask himself whether he would feel comfortable explaining his actions in any subsequent arbitration.

What architects and other consultants find difficult to accept is that they have no duty to the employer in these matters. Their professional skills are at the service of the contractor. Another situation which frequently arises is when the architect has detailed certain things on his drawings and possibly in a specification which he has produced for the contractor to enable construction to proceed on site. The contractor may indicate to the architect that he fully intends to construct the detail by another method and the architect may consider the contractor's method vastly inferior to the one he has detailed. The architect certainly has a duty to point out to the contractor that he seriously disagrees and his reasons for so doing, but if the contractor insists on doing things his own way or even ignores the architect's letter, the architect has discharged his duty and he can do no more. Of course, any consultant who is regularly ignored may decide that he no longer wishes to work under these circumstances. Whether he is entitled to terminate his employment for that reason depends on the terms of the agreement.

These problems are fairly simple to resolve provided the consultant remembers that he is preparing design and construction information for the benefit of the contractor, not for the benefit of the employer. Such an idea frequently goes against the grain for consultants unused to working for contractors, but in fact there is nothing unprofessional or unlawful in it. The employer may have some of the design produced by his own architect as part of the Employer's Requirements. In such a case, it seems that the architect engaged by the contractor would have a duty to satisfy himself that the initial design was workable. He is not simply entitled to accept the initial design as correct. Doubtless, the contractor will expect his consultant to identify such shortcomings before the Contractor's Proposals are formulated if the consultant is engaged at that point. Unless there is something clearly wrong with the initial design or the Requirements, it is very unlikely that the consultant has any duty to simply point out areas where the design might be improved.

Of course the consultant, as a member of the human race, will owe some general duties in tort to the employer and to others. He may be liable in tort of negligence if his design results in injury or death to any person or if it causes damage to property other than the thing he has designed. He owes this duty to anyone who might foreseeably be affected by his design: *Murphy* v. *Brentwood District Council* (1990).

If the consultant is engaged by the employer to produce the first part of the design including the Employer's Requirements and probably application for planning permission, his liability to the employer is exactly the same as the liability of any consultant performing a partial service for a client. His duties should be set out in his conditions of engagement. In general, he must use reasonable skill and care in taking instructions from the employer and in expressing those instructions in the form of Employer's Requirements. They should be so framed that a contractor tendering on the basis of satisfying those Requirements will produce the kind of Proposals envisaged by the employer. This is no mean task and the consultant must take great care in recording the employer's brief in the first instance. A consultant who fails to draft the Employer's Requirements with sufficient care may be liable to the employer for functional inadequacy in the resultant building if sufficient link between drafting and inadequacy can be established.

If the consultant is engaged by the employer in the third instance, to deal with tendering, contract documentation and to act as agent under the contract, his duties will depend on the terms of engagement. There will normally be a duty to take reasonable skill and care in performing those duties. While acting as employer's agent, the consultant will be governed by the normal law of agency (see section 5.1). The contract gives the agent power to act for the employer for the receiving or issuing of applications, consents, instructions, notices, requests or statements and otherwise to act for him under any of the clauses; that is, unless the employer specifies to the contrary by written notice to the contractor. A consultant filling this role is in a completely different position to the architect under JCT 98. The agent has no duty to the employer to act fairly between the parties, neither does he owe such a duty directly to the contractor. Thus in no sense is the consultant fulfilling an independent function. The wording of the contract makes the position clear: *J.F. Finnegan Ltd* v. *Ford Seller Morris Developments Ltd (No 1)* (1991). There is no provision for any form of certification or other discretionary activity by the agent. Indeed, except for two clauses (clauses 5.4 and 11) the employer's agent is not specifically mentioned in the conditions.

A consultant must be scrupulous in acting within the authority given to him by the employer under the contract. It is quite common for an employer to wish to reserve certain functions to himself and in some instances the agent may find himself with few powers under the contract. A difficult situation may arise if the consultant acting as agent during operations on site is not the consultant who assisted the employer in the formulation of the Employer's Requirements. The second consultant may disagree with parts of the Requirements, more particularly if an initial sketch design forms part of the Requirements. Clearly, he must advise the employer if his objections to the Requirements are serious. Ideally, an employer wishing to engage a consultant purely for the purpose of obtaining tenders and acting as agent during the progress of the work should give the prospective consultant an opportunity to examine the Employer's Requirements before accepting the commission.

3.4 *Consultant switch and novation*

Something has already been said about this topic in section 1.3 and there is further information in section 6.1. An interesting liability situation may arise. There is no doubt that, in theory, the system promotes a smooth design process, because it simply continues with the same design team involved. But, as can be seen from section 3.3 above, while in contract with the employer, the consultants owe a duty to the employer to take reasonable skill and care in preparing the design and in giving advice in the best interests of the employer. After the design team enter into contracts with the contractor, their duty in respect of design and any related advice is owed to the contractor in the context of the contractor's reasonable profit expectations. Thus, the consultant has the same design obligation, but owed to different parties at the two stages. In addition, there are different obligations to advise during the stages. The danger is that the employer, and sometimes the consultant, will forget that in the second stage, the consultant owes no advisory duty to the employer. Thus, if the contractor instructs the consultant to change part of the design, the consultant must comply, because he is now acting for the contractor, even if the consultant knows or thinks he knows that the employer does not want that particular change. This is because the contractor has merely sublet the design to the consultant and as between the employer and the contractor, the contractor has the design responsibility.

The consultant has generally only contracted to carry out the contractor's instructions regarding the design. These instructions may well be that the consultant must complete the design in accordance with the Employer's Requirements, but are not necessarily so. A consultant's long-established client may find it hard to accept that the consultant is no longer looking after his best interests. Each project demands individual consideration and often expert advice if consultant switch is contemplated. The consultant who is asked to become involved in design and build on the basis of consultant switch should take some time to explain these points to the employer before accepting the commission. Indeed, since the first edition of this book was written, anecdotal evidence suggests that consultants acting first for the employer and then for the contractor encounter considerable difficulties and our advice to consultants and to employers is to avoid these situations and act for one or the other party exclusively.

It also goes without saying, or should do, that a consultant engaged by the employer to prepare the Employer's Requirements should not take an engagement to work for a contractor until after the tendering process is complete, nor should there even be a tacit understanding. To do otherwise opens the consultant to the charge of partiality when the contractors invited to tender pose the inevitable questions.

From the employer's point of view, consultant switch can ensure a continuity of design, therefore less possibility of mistakes. It is not unusual, although inadvisable, for an architect to carry out initial design for the employer and to subsequently complete that design for the contractor while at the same time carrying out duties as employer's agent. It need hardly be said that such an arrangement will almost certainly give rise to a conflict of interest so far as the architect is concerned. It is rarely possible for the architect to keep such interests apart. A further danger is that the employer might be encouraged to think of the architect as his architect under a traditional procurement situation rather than as agent with limited authority and duties. For example, when acting for the contractor in completing the design drawings, etc., the architect's duties as agent do not extend to informing the employer of instructions received from the contractor to amend certain parts of the design. Neither can the architect take instructions regarding the design directly from the employer. In such a situation, the employer must instruct the contractor who, in turn, should instruct the architect. If the contractor chooses not to instruct the architect, the architect may not carry out the employer's instructions, even though he knows that they have been given. The contractor may have many reasons of his own for deciding not to give instructions to the architect. Clearly, architects should not agree to act for both parties in this way.

The system whereby a consultant acts first for the employer and later for the contractor is commonly, and often mistakenly, called 'novation'. Whereas consultant switch involves the consultant entering into terms of engagement with the employer and then into a completely different contract with the contractor when the first comes to an end, novation occurs where a contract between employer and consultant is replaced by a contract between contractor and consultant on identical terms. It is perhaps easier, although inaccurate, to visualise novation as taking away one party to a contract and replacing with another. Novation requires an agreement between all three parties, but the major problem is that the contractor will not want the same terms as the employer (for example, he will not require the same services). Therefore, if it is to be effective, the novation must make provision for a change in the terms. The benefit of novation is supposed to be that the consultant is made liable for all the design, even for early design carried out directly for the employer, and that this liability is owed to the contractor. Unfortunately, the protagonists of novation forget that the contractor is only liable under WCD 98 for *completing* the design, irrespective of whether the consultant is liable to him for the whole design. Therefore, the employer's attempt to channel all design responsibility through the contractor will fail unless WCD 98 itself is fundamentally amended.

3.5 *In-house or sublet*

From a liability standpoint, it makes little difference whether the contractor's design input comes from his own in-house design department or from independent consultants especially engaged for the project or from a combination of the two. There are a few practical points to note, however. A contractor who engages independent consultants must take care that his terms of engagement with them are back to back with his liabilities under the main contract WCD 98. It is not just sufficient to refer to WCD 98 in the terms of engagement. Even where no amendments have been made to the main contract clauses, there will always be the variable parts of the contract, the articles and the appendices, of which the consultant must have knowledge when entering into an agreement with the contractor. In many cases there will also be amendments to the contract itself and the consultant must have reference to all the terms of the main contract in his own terms. A particular point which the contractor should watch concerns professional indemnity insurance. It is essential that each consultant has insurance appropriate to the risks to be undertaken bearing in mind the terms of the main contract. Ideally, each consultant should provide the contractor with documentary evidence to this effect from the insurers.

Where design is to be carried out in-house, the contractor has vicarious liability for the actions of his employees. An architect or engineer in this situation is liable to the contractor for negligence in precisely the same way as any other employee. The fact that, in practice, professionals in employment do not carry personal professional indemnity insurance will probably ensure that it is not worthwhile for the contractor to sue them. In such cases, the contractor will have to carry his own professional indemnity insurance. The professional employees will want to know that there is a waiver of the insurer's subrogation rights.

3.6 *Warranties*

Section 3.1.1 above set out the basic principles of liability. It is clear that a designer's liability in tort for negligence may be severely restricted. In order to overcome the problem, it is common for the employer to ask for collateral warranties (or duty of care agreements) from anyone with whom he is not in direct contract. In theory, such warranties should not be generally required in the case of a design and build contract, because all liability is gathered under the contractor's wing. There is no JCT standard warranty form produced with WCD 98, probably for this reason. However, part of the design may have been done for the employer and incorporated in the Employer's Requirements, and the contractor has no liability for it. In addition, it is always possible that the contractor may go into liquidation. In these instances, it is in the employer's interests to enter into collateral agreements with all the consultants and with any other subcontractors. The terms of the warranties will normally provide, at the very least, that the warrantor will use reasonable skill and care in carrying out any design function, selection of materials and the satisfaction of any performance specification. Where consultant switch is operated, separate warranties are a great advantage to the employer, because any design fault is the responsibility of the consultant whether engaged directly by the employer or through the medium of the contractor. This is not the place to discuss the wording or content of warranty forms in detail and the parties should obtain expert advice. Matters commonly covered by warranties include the following:

- Design
- Selection of materials
- Satisfaction of performance specification
- Professional indemnity insurance
- Assignment
- Copyright
- Deleterious materials
- Notice of termination in prospect and provision for novation
- Dispute resolution.

It remains to be seen whether collateral warranties are really effective or necessary. In the case of *Beoco Ltd* v. *Alfa Laval Co Ltd* (1992) a subcontractor's collateral warranty was held to impose liability for all loss in respect of a weld badly done by the defendant subcontractors, although on the facts the defective weld was not the cause of the damage.

CHAPTER FOUR
THE CONTRACTOR'S OBLIGATIONS

4.1 *Express and implied terms*

The contractor will have obligations in respect of both express and implied terms. This is something which is seldom appreciated by employers or contractors who tend to consider that the contract documents represent the whole of the terms governing the agreement. An express term is one which the parties have agreed; in contrast an implied term is one which is written into the contract as a matter of law. The doctrine of implied terms is of the greatest practical importance.

4.2 *Implied terms*

An implied term is one which was not expressed in writing or orally at the time the contract was entered into, but which will be implied into contracts in a number of instances:

- By statute: for example, the Sale of Goods Acts 1979 and 1994 and the Supply of Goods and Services Act 1982 set out several terms which will be implied into contracts which fall within the scope of those Acts.
- By custom or trade usage (*Symonds* v. *Lloyd* (1859)).
- By common law. In *Liverpool City Council* v. *Irwin* (1977) the House of Lords noted two distinct circumstances in which the courts might imply terms:
  - If it was necessary to give business efficacy to a contract.
  - If the term was simply spelling out what both parties knew was part of the bargain.
- Another ground for implying a term was stated in *Mackay* v. *Dick* (1881), namely 'where in a written contract it appears that both parties have agreed that something shall be done, which cannot be done unless both concur in doing it, the construction of the contract is that each agrees to do all that is necessary to be done

on his part for the carrying out of that thing, though there may be no express words to that effect'.

Even in a very detailed contract such as WCD 98, it is very likely that it will be necessary to imply terms, as happened in the case of *London Borough of Merton* v. *Leach Ltd* (1985) in a contract in the then current JCT 63 form.

The courts will not do so to improve a contract, but only if it is essential to enable the contract to work. The following are terms which are commonly implied into building contracts:

- The contractor will carry out his work in a good and workmanlike manner – that is, he will carry out the work using the same degree of skill as would an averagely competent contractor who is experienced in that kind of work: *Hancock and Others* v. *B.W. Brazier (Anerley) Ltd* (1966).
- The contractor will supply good and proper materials – that is, materials which are satisfactory for their purpose: *Young & Marten* v. *McManus Childs Ltd* (1969).
- The contractor undertakes that a building will be reasonably fit for its purpose if that purpose has been made known to him and if there is no other designer involved: *Viking Grain Storage Ltd* v. *T.H. White Installations Ltd* (1985).
- The contractor will complete the Works by the date for completion or, if no date is stated in the contract, within a reasonable time of being given sufficient possession of the site: *Fernbrook Trading* v. *Taggart* (1979).

These terms may be modified or superseded by the express terms of the contract which are normally given preference. In the traditional form of contract on JCT 98 terms where an independent architect has been retained to prepare detailed drawings and specification and to administer the contract during operations on site, it has been found necessary to imply terms to make the contract commercially effective.

A useful discussion of the implication of terms in a specially drafted 'design and construct' turnkey contract for the provision of a floating production and storage facility in a North Sea oil field is to be found in the case of *Davy Offshore Ltd* v. *Emerald Field Contracting Ltd* (1992).

In contracts based on WCD 98, however, the situation is likely to be somewhat different. There may well be significant gaps in specification between the essentially performance specification part of the Employer's Requirements and the perhaps less than exhaustive treatment of the operational specification part of the Contractor's Proposals. In such instances, it is clear that the law will imply appropriate terms to deal with workmanship and materials. The question of fitness for purpose is dealt with in detail in Chapter 3. Suffice it to say here that a term of this sort will be superseded by the provisions of clause 2.5. The contractor's powers and duties are listed in Figs 4.1 and 4.2 respectively. The remainder of this chapter deals with the most important of the contractor's duties.

4.3 *Express terms*

4.3.1 Contractor's obligations

The contractor's obligations are set out in what are probably the most important provisions of the contract:

- Article 1
- Clause 2.1

Article 1 is set out in full below; the text of clause 2.1 can be found in section 3.2:

> 'Article 1
> Upon and subject to the Conditions and, where so stated in Appendix 1, upon and subject to the Supplementary Provisions issued February 1988 which modify the aforesaid Conditions, the Contractor will, for the consideration mentioned in Article 2, both complete the design for the Works and carry out and complete the construction of the Works.'

In order to properly understand those obligations the definition of 'Works' contained in clause 1.3 should be studied:

> 'the works briefly described in the First Recital and referred to in the Employer's Requirements and the Contractor's Proposals and including any changes made to those works in accordance with this Contract.'

There is a slight inconsistency in what the contractor is required to construct between article 1 and clause 2.1. The former encompasses the description in recital 1 together with the reference in the

Fig. 4.1
Contractor's powers.

Clause	*Power*	*Comments*
4.2	Request the employer to specify in writing which contract provision empowers the issue of an instruction.	
5.1	Inspect the Employer's Requirements and the Contractor's Proposals.	These documents are kept by the employer.
8.1	Substitute materials, goods and workmanship for those described in the Employer's Requirements or the Contractor's Proposals or specification.	Only if the original goods, etc. are not procurable and the employer consents in writing.
12.2.1	Consent to a change in the Employer's Requirements which is or makes necessary any alteration or modification in the design of the Works.	Consent must not be unreasonably delayed or withheld.
12.4.1	Agree with the employer the valuation of changes and provisional sum work.	
12.4.2,A1	Submit a price statement.	Within 21 days of the latest of: • receipt of an instruction; *or* • receipt of sufficient information.
12.4.2,A5	Refer the price statement as a dispute or difference to the adjudicator.	On or after the 21 day period to which paragraph A2 refers if no notification has been given.
17.1	Consent to the employer taking partial possession before practical completion.	
18.1.1	Assign the contract.	With the written consent of the employer.
18.2.1	Sublet all or part of the Works.	With the written consent of the employer.

Fig. 4.1 *Contd*

Clause	*Power*	*Comments*
18.2.3	Sublet all or any part of the design.	With the written consent of the employer.
22B.2	Require the employer to produce for inspection the insurance policy and the premium receipts. Insure in joint names all work executed, etc. against all risks.	Where the Works are to be insured in joint names by the employer. If the employer fails to produce receipt on request.
22C.3	Request employer to produce receipt showing that he has an effective policy under clauses 22C.1 and 22C.2. Insure in the employer's name and on his behalf and for that purpose enter the premises to make an inventory and survey.	Applies to existing structures where employer is to insure in joint names. If the employer fails to produce the premium receipt when requested unless the employer is a local authority.
22C.4.3.1	Determine his employment under the contract if it is just and equitable to do so. Invoke dispute resolution procedures.	This must be done within 28 days of the occurrence of the loss or damage, and is effected by written notice to that effect served on the employer. If the employer serves notice determining the contractor's employment; *and* the contractor alleges that it is not just and equitable to do so; *and* the contractor acts within seven days of receiving the notice.
22D.1	Require information from the employer.	To obtain a quotation under this clause.
23.3.2	Consent to the employer using or occupying the site of the Works before the issue of the practical completion certificate.	If insurers confirm that insurance will not be prejudiced, consent must not be unreasonably withheld.

Fig. 4.1 *Contd*

Clause	*Power*	*Comments*
26.1	Write to the employer stating that he has incurred or is likely to incur direct loss and/or expense and quantify the same, not reimbursable under any other contract provision.	If it becomes or should have become reasonably apparent that regular progress has been and is likely to be affected by one or more of the specified matters.
28.2.1	Give notice to the employer specifying the default.	If employer: • fails to pay by final date for payment; *or* • fails to comply with assignment provisions; *or* • fails to comply with the CDM Regulations.
28.2.2	Give notice to the employer specifying the suspension events.	If the Works are suspended for a period stated in appendix 1 due to: • late instructions, etc.; *or* • certain instructions; *or* • employer's operatives; *or* • failure to give access.
28.2.3	Give notice determining his employment.	If the default or suspension is not ended within 14 days.
28.2.4	Give notice determining his employment.	If the default or suspension event is repeated.
28.3.3	Give notice determining his employment.	If the employer becomes insolvent.
28A.1.1	Give seven day notice of suspension events	If the Works are suspended for a period stated in appendix 1 due to: • *force majeure*; *or* • loss by specified perils; *or* • civil commotion; *or* • development control requirements; *or* • certain employer's instructions; *or* • hostilities; *or* • terrorist activity.

Fig. 4.1 *Contd*

Clause	*Power*	*Comments*
29.2	Consent to the carrying out of such work by others.	If the employer so requests where the Employer's Requirements do not so provide. Consent must not be unreasonably withheld.
30.3.8	Suspend performance of obligations.	If the employer fails to pay in full by the final date for payment and the contractor has given a seven day notice of intention to suspend. The contractor cannot suspend if the employer has properly served a notice under clause 30.3.4.
30.4.2.2	Request the employer to place retention in a separate bank account.	
39A.3	Agree on someone to replace the adjudicator or apply to the nominator.	If he cannot act.
39A.5.2	Send to the adjudicator a written statement of the contentions relied on and any material he wishes the adjudicator to consider.	If not making the referral. Must act within seven days of the referral date with a copy to the employer.
39A.7.3	Take legal proceedings to secure compliance.	If the employer does not comply with the adjudicator's decision.
39B.1.3	Give a further notice of arbitration to the employer and to the arbitrator referring another dispute which falls under article 6A.	After an arbitrator has been appointed. Rule 3.3 applies.
S6.2.2	Raise reasonable objection to provision of estimates within 10 days of receipt of instruction.	May do so on behalf of any subcontractor.
Fluctuation clauses are not covered in this table.		

Fig. 4.2
Contractor's duties.

Clause	*Duties*	*Comments*
2.1	Carry out and complete the Works referred to in the Employer's Requirements, the Contractor's Proposals and the other contract documents and in accordance therewith complete the design for the Works including the selection of any specifications for any kinds and standards of materials, goods and workmanship so far as not described or stated in the Employer's Requirements or Contractor's Proposals. Comply with any instruction and be bound by employer's decisions under the contract.	Unless varied under clauses 39A, 39B or 39C.
2.3	Give the employer written notice immediately on finding any divergence between the Employer's Requirements and the definition of the site boundary.	
2.4.1	Inform the employer in writing of proposals to deal with the discrepancy.	If Contractor's Proposals do not deal with a discrepancy within the Employer's Requirements.
2.4.2	Inform the employer in writing of his proposed amendment where there is a discrepancy within the Contractor's Proposals.	
2.4.3	Give the employer written notice immediately, specifying any discrepancy discovered in the documents.	

Fig. 4.2 *Contd*

Clause	*Duties*	*Comments*
2.5	Design the Works using the same standard of skill and care as would an architect or other appropriate professional advisor holding himself out as competent to take on work for such a design.	
4	Forthwith comply with all instructions issued by the employer.	The instructions must be in writing *and* expressly empowered by the contract. The contractor need not comply where the instruction requires a change under clause 12.12 to the extent that he makes reasonable objection in writing to the employer.
4.3.2	Confirm to the employer in writing any oral instruction he issues.	
5.3	Provide the employer free of charge with two copies of the drawings, specifications, details, levels and setting out dimensions which he prepares or uses for the Works.	
5.4	Keep available to the employer's agent at all reasonable times one copy of the Employer's Requirements, Contract Sum Analysis, Contractor's Proposals and documents referred to in clause 5.3	
5.5	Supply the employer with 'as built' drawings free of charge.	This must be done before the defects liability period begins.

Fig. 4.2 *Contd*

Clause	*Duties*	*Comments*
6.1.1.2	Comply with all statutory requirements, give all notices, etc. and pass statutory approvals to the employer when they are received.	This includes making planning and related applications.
6.1.2	Notify the employer immediately in writing on finding any divergence between the statutory requirements and the Employer's Requirements or the Contractor's Proposals. Inform the employer in writing of his proposed amendment for removing the divergence. Complete at his own cost the design and construction of the work in accordance with the amendment.	With the employer's consent, which must not be unreasonably delayed or withheld. This is subject to clause 6.3.
6.1.3	Supply such limited materials and execute such limited work as are reasonably necessary to secure immediate compliance with statutory requirements. Forthwith inform the employer of the emergency and steps he is taking under clause 6.1.3.1.	In an emergency and if this is necessary before receiving the employer's consent.
6.2	Pay all statutory fees and charges and indemnify the employer against liability in respect of them.	No adjustment is made to the contract sum unless such fees are stated as a provisional sum in the Employer's Requirements.
6A.2	Comply with all the duties of a planning supervisor under the CDM Regulations.	While the contractor is and remains the planning supervisor.
6A.3	Comply with all the duties of a principal contractor under the CDM Regulations.	While the contractor is and remains the principal contractor.

Fig. 4.2 *Contd*

Clause	*Duties*	*Comments*
	Ensure that the health and safety plan has the features required by regulation 15(4). Notify any amendment to the plan to the employer.	
6A.4	Comply with all the reasonable requirements of the principal contractor.	Where the employer appoints a successor to the contractor as principal contractor and where the requirements are necessary for compliance with the CDM Regulations. Free of charge to the employer.
6A.5.1	Prepare and deliver the health and safety file under the CDM Regulations.	While the contractor remains the planning supervisor. Free of charge to the employer.
6A.5.2	Provide and ensure any subcontractor provides information for the preparation of the health and safety file.	Where the contractor ceases to be the planning supervisor.
8.1.1	Provide materials and goods of standards described in Employer's Requirements or Contractor's Proposals.	So far as procurable. If not so described.
8.1.2	Provide workmanship of standards described in the Employer's Requirements or Contractor's Proposals or appropriate to the Works.	To extent not so described.
8.1.3	Carry out all work in a proper and workmanlike manner and in accordance with health and safety plan.	
8.2	Provide the employer with vouchers to prove that the goods comply with clause 8.1.1.	If he so requests.

Fig. 4.2 *Contd*

Clause	*Duties*	*Comments*
8.3	Open up for inspection any work covered up and arrange for the testing of any materials or goods or executed work.	If the employer so instructs. The cost will be added to the contract sum unless the tests, etc. are adverse.
8.4	Remove from site work, materials and goods not in accordance with the contract.	If the employer so instructs.
8.6	Provide samples of goods and workmanship as specifically referred to in the Employer's Requirements or the Contractor's Proposals.	Before carrying out work or ordering goods.
9.1	Indemnify the employer against liability in respect of copyright, royalties and patent rights.	If the employer instructs the use of patented articles, etc. the contractor is not liable in respect of infringement, and all royalties, damages, etc. are added to the contract sum (clause 9.2).
10	Constantly keep a competent person in charge on site.	
11	Allow the employer's agent and any person authorised by the employer access to the Works, workshops, etc. at all reasonable times and ensure a similar right of access in any subcontract.	Subject to reasonable restrictions and protection of proprietary rights.
12.2.2	Notify the employer in writing whether he has any objection under regulation 14 of the CDM Regulations.	If the contractor is the planning supervisor, within a reasonable time after receipt of a change instruction.
12.4.2,A4.2	State whether or not he accepts the amended price statement.	Within 14 days of receipt of the amended price statement. If accepted, paragraph A3 applies.

Fig. 4.2 *Contd*

Clause	*Duties*	*Comments*
16.2	Make good at his own cost all defects, shrinkages and other faults specified in the schedule of defects.	The defect, etc. must be due to the contractor's failure to comply with his contractual obligations or to frost occurring before practical completion. The employer's schedule of defects must be delivered to the contractor not later than 14 days after the expiry of the defects liability period. The contractor must remedy the defects, etc. within a reasonable time of receipt of the schedule.
16.3	Comply with any instruction issued by the employer requiring the remedying of defects, shrinkages and other faults.	This is in addition to clause 16.2. No such instruction can be issued after the delivery of the schedule of defects or after 14 days from the expiry of the defects liability period.
17.1	Issue a written statement to the employer identifying the part of the Works taken into possession and giving the date of possession.	If employer with contractor's consent takes possession of part of the Works.
20.1	Indemnify the employer against any expense, liability, loss, claim or proceedings whatsoever in respect of personal injury to or death of any person.	If the claim arises out of or in the course of or is caused by the carrying out of the Works *unless* and to the extent that the claim is due to any act or neglect of the employer or of any person for whom he is responsible.

Fig. 4.2 *Contd*

Clause	*Duties*	*Comments*
20.2	Indemnify the employer against any expense, liability, loss, claim or proceedings whatsoever in respect of injury or damage to any property.	In so far as the injury or damage arises out of or in the course of or by reason of the carrying out of the Works; *and* is due to the negligence, omission or default of the contractor or any person engaged on the Works (other than the employer's people) or of their respective servants or agents. Loss or damage by specified perils to property with clause 22C.1 insurance is excluded.
21.1.1.1	Maintain necessary insurances for injury to persons or property.	The obligation to maintain insurance is without prejudice to the contractor's liability to indemnify the employer.
21.1.2	Produce documentary evidence of insurance cover.	When reasonably required to do so by the employer who may (but not unreasonably or vexatiously) require production of the policy or policies and premium receipts.
21.2	Maintain in joint names of the employer and the contractor insurances for such amount of indemnity as is stated in appendix 1 for damage to property other than the Works caused by collapse, subsidence, etc. Deposit with the employer the policy(ies) and premium receipts.	Where it is stated in the Employer's Requirements that this insurance is required. Employer must approve insurers.
22.3	Ensure that the joint names policies referred to in clause 2A.1 or clause 22A.3 *either*	If clause 22A applies.

Fig. 4.2 *Contd*

Clause	*Duties*	*Comments*
	• provide for recognition of each nominated subcontractor as insured; *or* • include insurer's waiver of rights of subrogation.	In respect of specified perils.
22A.1	Insure the Works in joint names against all risks for their full reinstatement value.	Applicable to new buildings – the obligation continues until the date of the statement of practical completion or date of determination. The insurance is to be placed with insurers approved by the employer. Clause 22A.3.1 enables this cover to be by means of the contractor's all risks policy.
22A.2	Deposit the policy(ies) and premium receipts with the employer.	
22A.4.1	Give written notice to the employer.	Forthwith on discovering loss or damage caused by risks covered by the joint names policy in clause 22A.1 or 22A.2 or 22A.3.
22A.4.3	With due diligence restore any work damaged, replace or repair any unfixed materials or goods that have been destroyed or damaged, remove and dispose of debris and proceed with the carrying out and completion of the Works.	After insurance claim under clauses 22A.1 or 22A.2 or 22A.3 and any inspection required by the insurers.
22A.4.4	Authorise insurers to pay insurance money to employer.	Acting also on behalf of subcontractors recognised pursuant to clause 22.3.
22B.3.1	Notify forthwith the employer of the extent, nature and location of any loss or damage	Upon discovering the loss or damage caused by risks covered by joint names policy in clause 22B.1.

Fig. 4.2 *Contd*

Clause	*Duties*	*Comments*
22B.3.3	With due diligence restore work damaged, replace or repair any unfixed materials or goods that have been destroyed or injured. Remove and dispose of debris and proceed with the carrying out and completion of the Works.	The restoration of damaged work, etc. is to be treated as a change in the Employer's Requirements and valued under clause 12.2.
22B.3.4	Authorise insurers to pay insurance money to employer.	Acting also on behalf of subcontractors recognised pursuant to clause 22.3.
22C.4	Notify forthwith the employer of the extent, nature and location of any loss or damage affecting the Works, etc.	The contractor must do this on discovering the loss, etc.
22C.4.2	Authorise insurers to pay insurance monies to employer.	Acting also on behalf of subcontractors recognised pursuant to clause 22.3.
22C.4.4.1	With due diligence reinstate and make good loss or damage and proceed with the carrying out and completion of the Works.	If no notice of determination is served or if the dispute resolution procedures, having being invoked, have decided against the notice of determination.
22D.1	Send quotation to the employer.	If the Employer's Requirements state and it is recorded in appendix 1 that liquidated damages insurance may be required and the employer has so instructed.
22FC.2.2	Comply with the Joint Fire Code and ensure his servants and agents or anyone else properly on site, other than employer's people and local authority or statutory undertakers, also comply.	

Fig. 4.2 *Contd*

Clause	*Duties*	*Comments*
22FC.3.1	Ensure remedial measures are carried out in accordance with employer's instructions by the remedial measures completion date.	If a breach of the code occurs and the insurers specify the remedial measures required and the time by which they must be carried out.
22FC.4	Indemnify the employer for breach of the code.	To the extent that they result from a breach by the contractor of his obligations under clause 22FC.
23.1.1	Begin the construction of the Works when given possession of the site. Regularly and diligently proceed with the Works and complete them on or before the completion date.	This is subject to the provision for extension of time in clause 25.
23.3.2	Notify insurers under clause 22A or 22B or 22C.2–.4 and obtain confirmation that use or occupation will not prejudice insurance.	Before giving consent to use or occupation.
24.2.1	Pay or allow the employer liquidated damages at the rate specified in appendix 1.	If the contractor fails to complete the Works by the completion date; *and* if the employer has notified him to that effect; *and* if the employer has given written notice not later than the date when the final account and final statement become conclusive that he may require payment of liquidated damages; *and* if the employer has given written notice requiring payment.

Fig. 4.2 *Contd*

Clause	*Duties*	*Comments*
25.2	Notify the employer in writing forthwith of the material circumstances (including the cause or causes of delay), identifying any event which in his opinion is a relevant event.	If and when it becomes reasonably apparent that the progress of the Works is or is being likely to be delayed. The duty is in respect of any cause of delay and is not confined to the relevant events specified in clause 25.4.
	Give particulars of the expected effects of any relevant event and estimate the extent, if any, of the expected delay to completion beyond the currently fixed completion date. Keep the particulars and estimate up to date by further written notices as may be reasonably necessary.	If practicable, this must be done in the above notice, but otherwise in writing as soon as possible thereafter.
25.3.4	Constantly use his best endeavours to prevent delay in progress to the Works and to prevent the completion of the Works being delayed or further delayed beyond the completion date; *and*	
	do all that may be reasonably required to the satisfaction of the employer to proceed with the Works.	The second part of this duty does not require the contractor to spend money.
26.1.2	Submit to the employer such information as is reasonably necessary to ascertain the amount of direct loss and/or expense.	Upon the employer's request.

Fig. 4.2 *Contd*

Clause	*Duties*	*Comments*
27.3.2	Immediately inform the employer in writing if he has made a composition or arrangement with creditors or, if a company, made a proposal for voluntary arrangement for composition of debts or scheme of arrangement.	
27.5.4	Allow and not hinder the taking of measures under this clause.	
27.6	Provide the employer with two copies of all drawings, details or descriptions, etc. as he has prepared or previously provided relating to the Works before determination of his employment.	In the event of determination by the employer under clauses 27.2.2, 27.2.3, 27.3.3, 27.3.4 or 27.4.
	Assign to the employer without payment the benefit of any subcontracts, etc. within 14 days.	Except where certain insolvency events have taken place.
	Remove from the Works any temporary buildings, plant, tools, equipment, goods and materials belonging to or hired to the contractor.	As and when so required by the employer in writing.
28.4	Provide the employer with two copies of all drawings, details or descriptions, etc. as he has prepared or previously provided relating to the Works before determination of his employment.	In the event of determination by the employer under clauses 28.2.3, 28.2.4 or 28.3.3.

Fig. 4.2 *Contd*

Clause	*Duties*	*Comments*
	With all reasonable dispatch and taking all necessary precautions to prevent injury, damage, etc. remove from site all temporary buildings, plant, tools, equipment, goods and materials and give facilities to his subcontractors to do the same. With reasonable dispatch prepare an account.	
28A.3	Provide the employer with two copies of all drawings, details or descriptions, etc. as he has prepared or previously provided relating to the Works before determination of his employment.	In the event of determination by the employer under clauses 28A.1.1.
28A.4	With all reasonable dispatch and taking all necessary precautions to prevent injury, damage, etc. remove from site all temporary buildings, plant, tools, equipment, goods and materials and give facilities to his subcontractors to do the same.	
28A.6	Provide the employer with all documents necessary for the preparation of the account.	Not later than two months after the date of determination under clause 28A.
29.1	Permit the execution of work not forming part of the contract to be carried out by the employer or by persons employed or otherwise engaged by him.	If the Employer's Requirements provide the contractor with such information as is necessary to enable him to carry out and complete the Works in accordance with the contract.

Fig. 4.2 *Contd*

Clause	*Duties*	*Comments*
30.3.1	Apply for interim payment, accompanied by such details as may be stated in the Employer's Requirements.	Under alternative A the applications are to be made on completion of each stage. Under alternative B applications are to be made at the period stated in appendix 2 up to and including the end of the period during which the day named in the statement of practical completion occurs and thereafter as and when further amounts are due. The employer is not required to make any such interim payment within one calendar month of having made a previous interim payment.
30.5.1	Submit to the employer the final account and final statement for the employer's agreement. Supply such supporting documents as the employer may reasonably require.	Within three months of practical completion.
31.3	Provide the employer with evidence that the contractor is entitled to be paid without statutory tax deduction or inform him in writing that he is not so entitled.	Where the employer is stated to be a 'contractor' for the purposes of the Income and Corporation Taxes Act 1988. The tax certificate must be provided not later than 21 days before the date of first payment under the contract.
31.4	Immediately inform the employer if he obtains a tax certificate. Provide the employer with evidence that he is entitled to be paid without statutory deduction or inform him that he is not so entitled.	Where the contractor is uncertified and then obtains a tax certificate. Where tax certificate expires before final payment. The evidence must be provided not later than 28 days before expiry of the certificate.

Fig. 4.2 *Contd*

Clause	*Duties*	*Comments*
	Immediately inform the employer in writing if the contractor's current tax certificate is cancelled, giving the date of cancellation.	
31.6	Indemnify the employer against loss and/or expense caused by an incorrect statement of direct cost to the contractor and any others of materials used in the carrying out of the Works.	If the contractor complies with clause 31.6.1.
34.1	Use his best endeavours not to disturb any fossils, antiques, etc. found on site and cease work if its continuance would endanger the object found or impede its excavation or removal. Take all necessary steps to preserve the object in the exact position and condition in which it was found. Inform the employer of the discovery and the precise location of the object.	
34.2	Permit the examination or removal of the object by a third party.	If the employer so instructs.
39A.4.1	Give notice to the other party of his intention. Refer the dispute to the adjudicator. Include particulars of the dispute and summary of contentions relied upon, statement of relief sought and any other material he wishes the adjudicator to consider.	Where the contractor requires a dispute to be referred to adjudication.
39A.5.7	Meet own costs of adjudication.	Except that the adjudicator may direct that a party bear the costs of tests or opening up if required.

Fig. 4.2 *Contd*

Clause	*Duties*	*Comments*
39B.1.1	Serve on the employer a notice of arbitration.	If the contractor wants a dispute referred to arbitration.
S2.1	Comply with provisions in Employer's Requirements re. submission of drawings and employer's rights to comment.	For this clause to apply, the provisions must be stated in the Employer's Requirements.
S3.2	Appoint a manager. Not to remove or replace the manager.	Prior to start of Works on site. The employer must have consented in writing. Without the employer's written consent which must not be unreasonably withheld or delayed.
S3.3	Ensure that the manager attends meetings in connection with the Works.	As and when reasonably requested by the employer.
S3.4	Ensure the manager keeps complete and accurate records and makes them available for the employer at all reasonable times.	In accordance with any provisions in the Employer's Requirements.
S4.2.1	Enter into contract with named subcontractor and notify employer of date.	
S4.2.2	Immediately inform the employer of the reasons.	If unable to enter into a subcontract with a named subcontractor.
S4.4.1	First obtain consent of the employer.	If he wishes to determine the named subcontractor's employment.
S4.4.2	Complete any balance of subcontractor's work.	If the named subcontractor's employment is left unfinished.
S4.4.3	Account to the employer for amounts recovered or which ought to have been recovered using reasonable diligence.	Reasonably due to the employer as a result of the determination.

Fig. 4.2 *Contd*

Clause	*Duties*	*Comments*
S4.4.4	Include a provision that the named subcontractor will not contend that the contractor has suffered no loss.	In the named subcontractor conditions.
S6.2	Within 14 days of an instruction, submit clause S6.3 estimates to the employer.	If compliance entails valuation, extension of time, or loss and/or expense. The contractor need not submit the estimates if the employer within 14 days of the instruction gives written statement that they are not necessary; *or* if within 10 days the contractor makes objection.
S6.4	Take all reasonable steps to agree the estimates with the employer.	After submission to the employer.
S7.2	Submit an estimate of the amount of loss and/or expense incurred in the period immediately preceding.	With the next application for payment if he is so entitled under clause 26.1.
S7.3	Submit an estimate of loss and/or expense with each application for payment.	For so long as he continues to incur loss and/or expense.
Fluctuation clauses are not covered in this table.		

Employer's Requirements and the Contractor's Proposals, while the latter seemingly deliberately omits the reference to the first recital where the description is probably written in very broad terms. It is not thought that the inconsistency will be a serious problem in practice. The contractor's obligation is absolute. There is no qualification. His obligation ends only at practical completion of the Works or upon the operation of the determination provisions.

The reference to changes was added by Amendment 6 in November 1990. It makes clear that reference to the 'Works' also includes all changes, in other words the Works in their final condition. On the face of it, the contractor is undertaking to carry out and complete the whole of the Works, including all changes, for the contract sum in article 2 and within the contract period stipulated in appendix 1. It is suggested, although the matter is not without doubt, that references to the contractor's obligations being subject to 'the Conditions' are sufficient to ensure that the provisions of clauses 12 and 25 are operated to provide the contractor with additional money and time as appropriate.

The principal difference between this and other contracts is of course the contractor's obligation to complete the design for the Works as well as carrying out and completing the construction of Works. Note that the obligation in respect of design is that the contractor must complete the design. For example, if the design was presented to him in the Employer's Requirements as fully completed, he would theoretically have no design obligation remaining. If the employer simply presented his Requirements in the form of a performance specification and nothing more, the contractor's design obligation would be total. The design obligation is discussed in detail in Chapter 3. Clause 2.1 is basic to the contract and often, when the parties are bogged down in argument regarding respective liabilities, it is worth returning to this clause which sets the rest of the contract in perspective. The second part of the clause is not as clear as it could be. After the obligation to 'complete the design of the Works', the remainder of the clause could mean either that it must not include the selection of any specifications for any kinds or standards of materials and workmanship if they are in the Employer's Requirements or the Contractor's Proposals, or it might mean that the design must be completed only to the extent that it is not in the Employer's Requirements or the Contractor's Proposals and that the specifications of standards of materials or workmanship are included in the design. On balance, we incline to the view that the second interpretation is correct.

4.3.2 Regularly and diligently

The contractor's duty to proceed regularly and diligently with the Works after being given possession of the site is to be found in clause 23.1. It is important, particularly in the context of the determination provisions. Failure to proceed regularly and diligently is ground for determination (see Chapter 12). It is discussed in detail in section 7.2.

4.3.3 Workmanship and materials

The contractor's obligations in respect of workmanship and materials are included in clause 8. They are similar to the obligations under JCT 98, but there are some significant differences. Clauses 8.1.1 and 8.1.2 deal with materials and goods and workmanship respectively. The contractor's duty is to provide materials and goods as described in the Employer's Requirements. If they are not described in the Requirements, the materials and goods are to be as described in the Contractor's Proposals or in the specifications referred to in clause 5.3, but only in so far as they are procurable.

This is a substantial protection for the contractor, because if it was not for this proviso he would have an absolute obligation to provide the goods and materials. That is to say, he would not be able to offer any excuse for failure to so provide. Of course, if this saving phrase was omitted, there would be situations arising in which the contractor would be unable to comply. In some circumstances, it may result in the contract being frustrated. In practice, his approach in those circumstances would be to suggest other materials for the employer's approval which are equal in every way to those specified at no additional cost to the employer and with no extension to the contract period. The effect of the saving provision is that if the contractor cannot obtain goods and materials as described, his obligations in this respect are at an end. This does not provide an escape for a contractor who has either delayed ordering until the goods are unobtainable or who is finding it more difficult than expected to get materials at a reasonable cost.

In order to decide whether this proviso applies in any particular situation, it is necessary to look at whether the contractor ordered the materials within a reasonable time after executing the contract, having regard to the date at which the materials were required on site. It cannot be expected that the contractor will immediately order every single item and store them on site or in his yard until required, but he should be alive to the possibility of materials becoming unobtainable. The provisions of clause 30.2B.1.2 are pertinent if interim payments are to be made. This provision refers to the value of materials which are not prematurely delivered to site. If the employer had the right to expect the contractor to order all materials at the very beginning of the contract, the contractor ought to have a complementary right to deliver to site and to receive payment. Neither would payment for materials off-site be discretionary. Clause 30.2A.4 is to similar effect with regard to stage payments where fluctuation clause 38 is applicable and certain other criteria are satisfied. If the materials are not procurable due to the fault of the contractor, it seems that he has an obligation to provide materials of an equivalent standard to the employer's satisfaction.

There is no precise machinery indicated in the contract when materials are not procurable. Under the traditional provisions of JCT 98, the materials which the contractor is unable to procure would be materials specified by the architect and an architect's instruction would be necessary to change the materials. Realistically, it is sensible for the employer to embody a substitute material in a change instruction so that it can be valued and the contract sum can be adjusted. In such circumstances, the contractor may be entitled to an extension of time and loss and/or expense. However, 'so far as procurable' does not give the contractor the discretion to amend materials as he thinks fit, even if the original materials are genuinely not procurable. Indeed, the clause further provides that the contractor may not substitute anything described unless he has the employer's written consent.

This provision clearly refers to a situation where the contractor, for reasons of his own, wishes to change the material from that described. The contract is silent regarding the mechanism to be employed if the contractor wishes to substitute. No doubt in many instances, the parties will deal with the situation in a fairly informal way. The contractor may well write to the employer suggesting the substitution of one material for another. An astute employer, realising that the contractor has made the suggestion because there is some advantage to be gained, may require the contractor to specify the saving in cost which the employer can expect if he agrees. However, there is no mechanism to achieve the reduction in the contract sum which should result. The only way to achieve that would be for the employer to issue an instruction requiring a change and so allow the substitution to be valued. Unfortunately for that line of approach, the contract does not seem to empower the employer to instruct a change in materials. Clause 12.1 refers to a 'change in the Employer's Requirements which makes necessary ...'. Among the things made necessary are the 'alteration of the kind or standard of any of the materials or goods to be used in the Works'. Clause 13 makes clear that the contract sum can only be adjusted or altered in accordance with the express provisions of the contract.

A further example highlights the problem. Is the employer entitled to a reduction in the cost of the Works if he simply writes and agrees to a drawing, submitted under clause 5.3 or clause S2, which shows the substitution of a cheap material for the more expensive one specified in the Employer's Requirements/Contractor's Proposals? At first sight, there appears to be no mechanism to deal with the situation and it might be said that if the employer simply agreed without precondition, he has agreed to the reduction in quality without a corresponding reduction in price. But, if there is no contractual procedure to accomplish a price reduction following a substitution, still less can there be the procedure to reduce quality *without* a price reduction. A more correct view may be that, whether or not the employer stipulates a price reduction before agreeing to the substitution, there would have to be a reduction (i.e. a payment from the contractor to the employer) as the consideration for the employer's agreement. Since there is no mechanism for adjusting the final account and final statement to reflect the reduction, the adjustment would have to be made outside the contract as part of a collateral contract. This kind of 'side-agreement' is relatively common in construction work, although often the parties are not aware of the precise legal relationship they are setting up and the variation is valued as though it had been *validly* instructed. That, of course, is the whole point. The obvious solution is to bring the situation within the change provisions so that it can be valued under the contract provisions. The matter is not without doubt and we hope that the JCT will direct their minds to the problem at the next amendment.

Under clause 8.2, if the employer so requires, the contractor must supply him with proof in the form of vouchers that the materials and goods comply with clause 8.1.1.

Clause 8.1.2 deals with workmanship. The standard of workmanship is to be as described in the Employer's Requirements. To the extent that the standard is not so described, it must conform with the Contractor's Proposals or the specifications issued by the contractor under clause 5.3. It is notable that in this clause and clause 8.1.1, noted above, the Employer's Requirements are given precedence. For example, if a material or standard of workmanship is variously described in the Requirements and the Proposals, it is clear that the description in the Requirements is to be preferred. Only if there is no specific description in the Employer's Requirements can the Contractor's Proposals be consulted. If no description is to be found in any document, the standard of workmanship must be appropriate to the Works. This is an imprecise yardstick, but most people in the construction industry will appreciate the difference in standards required for a cheap factory and a prestigious bank or office development.

If the parties cannot agree on the standard, it may be referred to adjudication under the provisions of clause 39A during the progress of the work, or to arbitration (see Chapter 13). In any such adjudication or arbitration, one of the factors to be taken into account may be the price being paid for the Works: *Cotton* v. *Wallis* (1955); *Phoenix Components Ltd* v. *Stanley Krett t/a North West Frontier* (1989). The situation should never arise in practice, because in the absence of an appropriate description, it is always open to the contractor to issue a specification note on the particular aspect of workmanship and to send a copy to the employer under the provisions of clause 5.3. Such specification will then be the standard prescribed under clause 8.1.2.

Clause 8.1.3 requires the contractor to carry out all work in a proper and workmanlike manner and in accordance with the health and safety plan (see also section 5.3.2). Whether a contractor is complying with the first part of this duty in any particular instance will be a matter to be decided with reference to any term in the Employer's Requirements or the Contractor's Proposals, any relevant code of practice and the practice in the industry.

The provision of samples is required by clause 8.6. However, this is by no means an all-embracing clause as is sometimes thought by employers and their agents. The provisions are quite precise and amount to the following: in respect of workmanship, goods or materials, if samples of the standards or quality of such workmanship, goods or materials are specifically referred to in either the Employer's Requirements or the Contractor's Proposals, and if the contractor intends to provide such workmanship, goods or materials, the contractor must provide the employer with such samples before either carrying out the work or ordering the materials.

For example, the Employer's Requirements may state that 'the contractor must provide a sample of the floor finish one metre square, on appropriate base to show the standard of workmanship and the quality of materials'. If it is the Contractor's Proposals, it will be rather more precise, because the contractor is essentially putting forward his specification for the work. On its true construction, it is considered that the clause must specify whether the sample is intended to show standard of workmanship or quality of materials or both. In some cases it will be obvious, for example where a sample of the type of brick or roof cladding is required. Not until the units are combined will any workmanship considerations apply. Generally worded admonitions in the Employer's Requirements, such as 'samples of all goods and materials intended for use on the Works must be provided to the employer before ordering', or 'samples must be provided to the employer as required from time to time', are not thought to be sufficiently specific to fall within the terms of this clause and they are void by operation of clause 2.2

giving priority to the printed form (see section 2.7). From a simple common sense point of view, the contractor will be unable to price for such vague provisions.

The contract is silent on the position if the employer dislikes the sample presented to him. There is no approval procedure and the contractor is entitled to provide the required samples and then simply proceed with the Works. The clause does not state that the contractor must provide the samples any particular time period before carrying out the work or ordering the materials and he would strictly comply with the clause if he provided them just the day before he was due to take action, although it may be prudent to allow a longer period. It is not thought that the courts would imply any term that a reasonable period should be allowed in view of the fact that where the contract wishes some action to be reasonable, it so states. It also specifies particular periods in other cases and the context of the contract as a whole does not require it: *R.M. Douglas Construction Ltd* v. *CED Building Services* (1985). This clause is clearly intended to enable the employer to check in certain circumstances as the work progresses that he is getting what the contract documents specify. But it will inevitably happen that the employer dislikes something he required, when he actually sees it. For this reason, the employer would be wise to amend the clause so that it stated a time period such as 'five working days', to give him the opportunity to decide whether he wishes to change his requirements. Of course, any such change would be subject to clause 12 (changes) and to clauses 25 (extensions) and 26 (loss and expense) if any delay or disruption was caused. The whole philosophy of this contract is to place responsibility on the contractor to satisfy the employer's carefully formulated requirements and, therefore, the employer should be wary of making expensive changes. This clause should be used for its primary purpose to check that the contractor is complying with the terms of the contract.

Clause 15.1 provides that the materials and goods which are intended for the Works and which are placed on or adjacent to the Works, but not fixed, may not be removed without the employer's consent in writing. The employer must not unreasonably withhold his consent. For example, it would be unreasonable if he withheld his consent to the removal of materials which for one reason or another were no longer required as part of the Works or which had been the subject of repeated thefts. The second part of this clause attempts to provide that the materials and goods will become the property of the employer after their value has been included in any interim payment. This clause will be effective against the contractor, but it cannot transfer the ownership of the materials to the employer unless, at the time of the purported transference, they belong to the contractor. This clause is not effective against subcontractors, sub-subcontractors or suppliers unless some similar provision is included in their contracts. This contract, unlike JCT 98, does not require the contractor to insert such clauses into his subcontracts, although there is a similar clause in the standard form of sub-contract DOM/2 published by the Construction Confederation (CC). Indeed, the provision would have to be inserted right down the contractual chain to overcome this problem: *Dawber Williamson Roofing Ltd* v. *Humberside County Council* (1979). Once the materials are incorporated into the Works, the ownership passes to the employer: *Reynolds* v. *Ashby* (1904).

Materials stored off-site by the contractor are dealt with in clauses 15.2 and 15.3. If the employer wishes to include the value of off-site materials in stage payments or in interim payments, he must have included them in a list attached to the Employer's Requirements. The contractor must satisfy the list of seven criteria. These criteria are intended to protect the employer against the risk of paying for materials which the contractor does not legally own and also to safeguard the employer if the contractor should become insolvent. In the latter situation, the employer can only recover the materials if he can show that there is no doubt that specific labelled materials belong to him (off-site materials are discussed at greater length in Chapter 10).

4.3.4 Statutory obligations

The obligations placed on the contractor by clause 6 are of the greatest importance in the context of his general design and build obligations. The key provision is clause 6.1.1.2 which states that the contractor must not only give all statutory notices, he must also comply with all Acts of Parliament, instruments, rules or orders or any regulation or byelaw of any local authority or statutory undertaker with any authority in regard to the Works or having systems (e.g. main sewers) into which the Works are or will be connected. This is expressly stated to include development control requirements which are defined in clause 1.3 as any statutory provisions and any decision of a relevant authority thereunder which controls the right to develop the site. In simple terms, the contractor must comply with the requirements for obtaining planning permission. The contractor must pass all approvals to the employer.

The contractor, therefore, must ensure that his Proposals comply with all local planning authority requirements and in this connection it is vital that the employer spells out in his Requirements just what is required. It is quite common for the employer to have obtained outline planning permission or even full planning permission while leaving the contractor to comply with conditions imposed by the planning authority. There is an important proviso in clause 6.1.1.1 that if the employer states in his Requirements that they comply with statutory requirements, the contractor has no duty under the contract to so comply nor to give any notices relating to the subject of the compliance. If the employer states that his Requirements comply partially, then to the extent that they comply, the contractor's obligations are reduced accordingly. Thus, the employer should make sure that copies of any applications and approvals he has sent and received are included as part of the Employer's Requirements. In practice, it is often unclear whether the employer has stated that his Requirements comply. In such circumstances, the contractor should seek written confirmation of the position at the time of formulating his Proposals before tendering.

Clause 6.3.2 provides that if the terms of any permission or approval of the planning authority after the base date stated in appendix 1 have the effect of amending the Contractor's Proposals, the amendment is to be treated as if it is the result of a change in the Employer's Requirements under clause 12.2. This situation could arise in several ways. The employer may obtain outline planning permission or even full planning permission with reserved matters and the contractor may produce his Proposals and tender on this basis. After the base date (which is normally a date about the date the tender is submitted), the planning authority may give decisions on the reserved matters which are in conflict with what the contractor has included in his Proposals. In such circumstances and provided the contractor has formulated his Proposals on the basis of information then available, he is entitled to have the necessary amendment to his Proposals treated as a change in the Employer's Requirements and reimbursement would follow. In another situation, the employer may not have obtained any planning permissions. When the contractor eventually receives permission, any amendments necessary to his Proposals will rank as if they were made necessary by a change to the Employer's Requirements. There is a proviso that the employer has not precluded such treatment in his Requirements. If he has so precluded, the contractor must stand the increased costs resulting from such amendments. An employer who seeks to place such total responsibility on the contractor will usually pay a heavy price in an increased tender figure.

If the employer has specifically stated in his Requirements that they or any part comply with statutory requirements, any amendments to his Requirements in order to conform with statutory requirements are dealt with under clause 6.3.3 and must be the subject of a specific instruction of the employer requiring a change. If it was not for this contractual provision, the employer's statements in his Requirements could be held to be misrepresentations and the contractor would have his remedies under common law or under the provisions of the Misrepresentation Act 1967. Changes in statutory requirements after base date requiring amendments to the Contractor's Proposals are covered by clause 6.3.1 and they are to be treated as an instruction from the employer under clause 12.2 requiring a change. The contractor is obliged to comply with statutory requirements even if the Employer's Requirements do not so comply. This must be the case, because the contractor's duty to comply with statutory requirements takes priority over his contractual obligation to satisfy the employer. In addition, clause 2.2 ensures that the requirements of clause 6 take precedence over the Employer's Requirements.

Clause 6.1.2 provides that if either the employer or the contractor finds a divergence between statutory requirements and the Employer's Requirements (including any change under clause 12.2) or the Contractor's Proposals, the finder must give immediate written notice to the other. Whoever gives the notice, the contractor must send the employer written proposals for removing the divergence. Provided the employer consents, and he may not unreasonably withhold such consent, the contractor must proceed to incorporate the amendment at his own cost. There is an important proviso that the amendment will not be at the contractor's cost if the problem is caused by one or more of the situations envisaged in clause 6.3 discussed above, which may be summarised as changes to statutory requirements after the base date or amendments necessary to the Employer's Requirements which are specifically stated to comply.

Clause 6.1.3 makes provision for any emergency compliance with statutory requirements. This will normally involve some danger to health and safety or imminent structural collapse, leaving no time for the contractor to seek specific instructions from the employer or his agent. The contractor must supply the minimum necessary materials and carry out the minimum amount of work necessary to comply and overcome the emergency. The extent to which the contractor is entitled to be reimbursed will depend on the precise circumstances in accordance with the various principles set out above. As soon as reasonably practicable, the contractor must let the employer know the problem and the steps being taken to overcome it.

Clause 6.2 deserves careful attention. As might be expected, it makes the contractor responsible for paying all fees and charges legally demandable under any Act of Parliament, etc. in connection with the Works. The contractor must include for all such payments in his tender figure. Only if they are expressed as provisional sums in the Employer's Requirements will he be entitled to receive the actual amount expended. Provisional sums are unlikely to be included for this purpose unless the figures are likely to be of such size and so unpredictable that the contractor will be obliged to include a very large sum in his tender figure to cover the risk. A point which is often overlooked is the indemnity which the contractor gives to the employer in respect of liability for the fees and charges. Its effect is that if the contractor fails to pay, he agrees to reimburse the employer, not only for the actual amount of the charges, but also for any consequential losses without restriction. Its effect is much broader than the employer's normal remedy for the contractor's failure which would be to sue for damages for breach of contract.

It is worthwhile considering the position of what are referred to in the contract as statutory undertakers. These are organisations such as the water supplier, gas suppliers and electricity suppliers which are authorised by statute to construct and to operate public utility undertakings. They derive their powers from statute either directly or through statutory instruments and the position is not changed for the purposes of this contract by the privatisation programme. They can be involved in the contract either in performance of their statutory obligations or as contractors or subcontractors. In the performance of their statutory obligations, they are not liable in contract: *Willmore* v. *S.E. Electricity Board* (1957), but in certain circumstances they may have a tortious liability. It is possible that they are directly engaged by the employer under clause 29 or they may be a subcontractor under clause 18.2.1 or clause S4 or they may be a statutory undertaker. It is important to establish in which capacity they are on site, because if they disrupt the regular progress of the work, the contractor will be entitled to an extension of time and loss and/or expense if they are considered to be employer's licensees under clause 29. He will be entitled to an extension of time if they are simply acting in their capacity of statutory undertakers. If acting as named or domestic subcontractors, the contractor must bear the risks of disruption himself unless he can recover his losses from the statutory undertakers.

Amendment 8 introduced provisions requiring the parties to comply with the CDM Regulations. By making compliance with the Regulations a contractual duty, breach of the Regulations becomes a breach of contract so providing both employer and contractor with remedies under the contract. Clause 6A.1 provides that the employer 'shall ensure' that, if the contractor is not the planning supervisor, the planning supervisor carries out all his duties under the regulations and that, where the principal contractor unusually is not the contractor, the principal contractor will also carry out his duties in accordance with the regulations. If the contractor is the planning supervisor, he must carry out all the duties of a planning supervisor under the regulations (clause 6A.2). There are also provisions that the contractor, if he is the principal contractor, will comply with the Regulations and 'ensure' that any subcontractor will provide the information to the principal contractor which the planning supervisor will reasonably require for the preparation of the health and safety file (clauses 6A.5.1 and 6A.5.2).

Every change instruction issued by the employer potentially carries a health and safety implication which must be examined and the appropriate procedural steps taken under the regulations. The regulations present the planning supervisor with very grave responsibilities. Some of his duties must be carried out before work is started on site. If necessary actions delay the issue of a change instruction or once issued delay its execution, the contract provides that the contractor is entitled to extension of time (clause 25.4.16) and any loss and/or expense he can substantiate (clause 26.2.8). Every construction professional should have a thorough grasp of the regulations and the contractual clauses which deal with them so that the full consequences of any new instruction can be carefully considered before it is issued.

4.3.5 Person-in-charge

Clause 10 requires the contractor to keep on site a competent person-in-charge. He must be on site 'constantly', i.e. during the whole of the time the Works are being executed. The contractor may designate anyone as person-in-charge and the idea is that there is always someone available on site to whom the employer can issue instructions confident that such instructions are being issued to the contractor. It is, therefore, essential that the person-in-charge thoroughly understands the implications of clause 4 and in particular, the need to get all instructions in writing before complying.

Supplementary provision S3 provides for the appointment of a site manager. Where these provisions apply, they replace clause 10. From the employer's point of view, it is worthwhile applying S3 because it requires the contractor to obtain the employer's written consent before appointing the site manager and in the case of any change in the appointment. The employer may not unreasonably withhold his consent to removal or replacement of the site manager, but it seems that there is no such restriction in respect of the initial appointment. The site manager is to be full-time on site and instructions given to him are deemed to have been given to the contractor. Clause S3.3 requires the manager, together with the contractor's servants, agents, suppliers or subcontractors as necessary, to attend any meetings which the employer may convene in connection with the Works. The manager is also required, by clause S3.4, to keep complete and accurate records in accordance with any provisions which are included in the Employer's Requirements. Therefore, if there are no provisions concerning records in the Employer's Requirements, the manager has no duty to keep such records under this clause. In practice, of course, any competent manager will keep records for his own and the contractor's benefit. Such records will contain details of weather, visitors to site, instructions given, deliveries, men employed and on which operations, progress, notable occurrences and so on. If the employer has specified particular records in his Requirements, the manager must make them available for inspection by the employer or his agent at all reasonable times, i.e. during normal working hours.

4.3.6 Instructions

Clause 4 is vital to the proper execution of the contract. It is discussed in detail in section 5.3. The contractor's principal duties under this clause are:

- He must comply with all instructions issued to him by the employer as soon as he reasonably can.
- This is subject to the contractor's right to query the empowering provision in the contract. The contractor may accept the employer's response and thereby receive all the benefits flowing

from the instruction whether or not it actually is empowered under the specified clause.

- The contractor need not comply if he makes reasonable objection to an instruction requiring the imposition of any obligations or restrictions or alterations to such obligations or restrictions in respect of access to the site, limitations of space or hours or the execution of the work in a specific order; *or*
- If the instruction results in an alteration to the design of the Works and with good reason, he does not consent; *or*
- If the contractor is the planning supervisor and he has an objection under regulation 14 of the CDM Regulations. If so, he must give written notification to the employer within a reasonable time of receiving the instruction. The employer must vary the instruction to remove the cause of the objection until the contractor is satisfied. The contractor may not continue his objection unreasonably; *or*
- If the instruction is oral. In this case the contractor must confirm it in writing to the employer within seven days. If the employer does not dissent, the contractor's obligation to comply takes effect from the expiry of seven days from receipt of the contractor's confirmation. The contractor need not confirm if the employer does so first. The contractor must comply from the date of the employer's confirmation. There is provision in clause 4.3.2.2 for the employer to confirm at any time before the conclusivity of the final account and final statement if neither party confirmed, but the contractor nevertheless complied with an oral instruction.
- If the contractor fails to comply with a written notice from the employer requiring compliance within seven days, clause 4.1.2 empowers the employer to engage others to carry out the instruction and charge all costs in connection with the operation to the contractor.

4.4 *Other obligations*

4.4.1 Drawings

The contractor's obligations to provide drawings are found in clauses 5.3 and 5.5. Clause 5.3 stipulates that the contractor must provide to the employer without further charge two copies of the drawings, specifications, details, levels and setting out dimensions which he prepares or uses for the Works. The drawings are for information only. It is clear from the express reference to setting out dimensions and levels that, despite the fact that the employer must define the site boundaries in accordance with clause 7, the contractor is responsible for setting out on site. That duty would in any event be implied as a vital part of the contractor's obligations under the design and build contract.

Clause 5.5 requires the contractor to supply the employer, before practical completion, with as-built drawings and other information showing operational and maintenance details. His obligation is not open-ended, but merely to supply such drawings and other information as may be specified in the 'documents named in clause 5.1'. This is a rather inelegant way of referring to the Employer's Requirements and the Contractor's Proposals.

Where the supplementary provisions apply, the contractor may have further obligations. Clause S2 assumes that the Employer's Requirements make provision for the submission of drawings to the employer. If there is no such provision, clause S2 will not work. If the employer wishes to keep a degree of control of the Works, he will make such provision. On the assumption that he does, clause S2.1.1 obliges the contractor to comply with whatever provisions the employer makes in regard to the submission of drawings and other details prior to their use on the Works. The drawings referred to are those specified in the Employer's Requirements or, if none are specified, such drawings, etc. as are reasonably necessary to explain or amplify the Requirements or the Proposals or to enable the contractor to carry out the Works or comply with any employer's instruction. The contractor must also comply with any provision concerning the employer's rights to comment on the information submitted.

Clause S2.2 stipulates that the contractor is not relieved from any liability by virtue of the employer having commented or not commented on the information unless, which seems unlikely, the comments specifically so state. Clause S2.3 notes that clauses 5.3 and 5.6 apply to the information referred to in this clause S2.

4.4.2 Copyright, royalties and patents

The contractor must include for all royalty payments, etc. which are payable in relation to any supply or use of anything in connection with the Works (clause 9). This will generally include everything expressed or inferred in the Requirements or the Proposals. In addition, the contractor indemnifies the employer against all claims which may be brought against him as a result of any infringement by the contractor of any patent rights or the like. The effect of this provision is that the contractor agrees to reimburse the employer for all costs in connection with such infringement without limitation.

If the contractor infringes any rights as a result of complying with the employer's instructions, any money which the contractor is liable to pay will be added to the contract sum as reimbursement to the contractor.

4.4.3 Access to the Works

The contractor is obliged to give access to the Works, and other places where work is being prepared, for the employer's agent and any person authorised by the employer. The contractor must also include terms in his subcontracts to achieve similar rights of access to subcontractors' workshops. The contractor must do everything reasonably necessary to make such rights effective. There is an important proviso that the contractor and any subcontractor may impose reasonable restrictions to safeguard their proprietary rights.

CHAPTER FIVE
THE EMPLOYER'S POWERS AND DUTIES

5.1 *Employer's agent*

The employer's agent is provided for in article 3. The choice of the term is deliberate and there is no sense in which the employer's agent is performing the same function as an architect under JCT 98. Under the traditional form of contract, the architect not only acts in an agency capacity, he also owes a duty to the employer to act fairly between the parties: *London Borough of Merton* v. *Stanley Hugh Leach Ltd* (1985). The employer's agent, though he may be an architect, is generally thought to have no such duty under WCD 98, although this is arguably not so. The Court of Appeal in *Balfour Beatty Civil Engineering Ltd* v. *Docklands Light Railway Ltd* (1996) said in relation to a different form of contract:

> 'We would ... have wished to consider whether an employer vested with the power ... to rule on his own and a contractor's rights and obligations, was not subject to a duty of good faith substantially more demanding than that customarily recognised in English contract law'.

The decision in this case has subsequently been questioned in *Beaufort Developments (NI) Limited* v. *Gilbert Ash NI Limited* (1998), but not on this point. There is little doubt, however, that the employer's agent under this contract does not have the same status in the eyes of a court or an arbitrator as if he was an independent architect engaged under a traditional contract.

At first sight, the last sentence of clause 2.1 appears to invest the employer with considerable powers. It states that the contractor is bound by any decision of the employer 'made under or pursuant to the Conditions'. However, the end of the sentence makes clear that the contractor is bound only to the extent that he does not challenge it in adjudication, arbitration or litigation as the case may be. This perfectly sensible provision requires that the employer has acted when obliged or empowered to do so in accordance with the contract. Only that kind of decision is binding. This is an altogether different concept to the provision in some contracts that the architect/engineer/contract administrator's decision is final and binding. That kind of provision would not be open to review in the dispute resolution procedure.

The principal (the employer) is bound by the properly authorised actions of his agent. It is important to establish the extent of the agent's authority. It may be actual or apparent (ostensible). The employer's agent's actual authority is defined by the terms of his agreement with the employer. His apparent authority may be quite different. Apparent authority is the authority the agent seems to possess as viewed by persons other than the employer. The agent's authority, so far as the contractor is concerned, will be laid down in the terms of the contract WCD 98 and in the Employer's Requirements. Therefore, if the authority given to the agent by WCD 98 is not matched by his actual authority under his agreement, the contractor is entitled to take account of his apparent authority and the employer will be responsible for the agent's actions pursuant to the provisions of WCD 98. The agent, however, having exceeded his actual authority, will be liable to the employer for the consequences. It is, therefore, crucially important that the employer's agent ensures that his agreement with the employer gives him at least the same degree of authority as apportioned to him in WCD 98 and the Employer's Requirements, and preferably rather more. In addition, the exact scope of that authority must be clearly defined.

An agent's duties to his principal are:

- To act. The employer may sue if his agent fails to act at the appropriate time.
- To obey instructions. The proviso is that the instructions must be lawful and also reasonable.
- To declare to the principal if there is any conflict of interest.
- To keep proper accounts.
- Not to take any secret bribe or profit. Breach of this duty entitles the principal to recover damages which may include the amount of the bribe or profit.
- Not to delegate without authority. This is important in the context of the duties which WCD 98 lays at the agent's door.

In the circumstances envisaged by WCD 98, the agency relationship will almost certainly be the subject of an express appointment. It should be in writing to reduce the possibilities of misunderstandings. There are, however, three other ways in which agency may arise, which we include here for the sake of completeness. If the employer acts to others as though a person is his agent and that person is acting like an agent, agency may arise by implication. If a person acts for another in an emergency, the agency may be created by necessity. Finally, it sometimes happens that a person performs some action for another and the other then ratifies it. Ratification requires two conditions to be satisfied: the agent must perform the action on behalf of the principal and the principal must have been capable of carrying out the action at the time it was performed.

Article 3 refers to the employer's agent as a 'person'. In normal principles of interpretation, it is not thought that the term need be taken to mean an individual. Indeed, in clause 1.3 of JCT 98, 'person' is defined as 'an individual, firm (partnership) or body corporate' although the word is not defined at all in this contract and the definition in JCT 98 has no relevance. The person nominated as agent may well be an employee of the employer. This is likely to be the case if the employer is a large corporation with its own in-house professional staff. In other cases, the agent will probably be an appropriate professional person such as an architect, an engineer or a surveyor. Where the person nominated is the head of an in-house department, it will be accepted that the duties of agent will be carried out by his staff and the same is true where a private consultant is named. On a strict reading of the provision this is not correct and, to be correct, the employer ought probably to name an individual. Article 3 clearly provides for the employer to nominate a replacement without any reason being given and there appear to be no formalities involved (other than the obvious necessity of written notice) and no right for the contractor to raise any objection. The person nominated as agent need not necessarily be a professional, nor need he be involved in construction. Clearly, however, the employer would be well advised to appoint a construction professional in this role.

The employer will often appoint an architect in the first instance to assist him in preparing the Employer's Requirements and any drawings required before the contract is executed, and to give general advice. Because of his knowledge of the job, it makes a great deal of sense to appoint that person as employer's agent unless, of course, a 'consultant switch' is to be operated (see section 3.4). Whoever is appointed as agent under this contract should be quite clear that the role is far removed from the role of an architect in a traditional contract. The reality is that many architects in this position proceed as they would under a traditional contract, issuing certificates and instructions, ascertaining loss and expense and so on. What is required is a completely different approach which should become clear in this chapter. There appears to be no reason why the employer should not have other advisors, such as quantity surveyors and engineers.

The employer's agent is only referred to in clauses 5.4 and 11 in relation to the availability of drawings and access to the site respectively. The employer's agent or any other person authorised by the employer or his agent is to have access at all reasonable times to the Works, workshops and other places where work is being prepared for the contract. The contractor is also to obtain similar access rights from subcontractors. The contractor and any subcontractor may impose whatever restrictions are necessary to safeguard their trade secrets. This clause is very similar to the equivalent clause in JCT 98.

The contract empowers the agent to act for the employer in receiving or issuing applications, consents, instructions, notices, requests or statements or for otherwise acting for the employer under any of the conditions. The article could scarcely be more broadly drafted. If the employer wishes some other arrangement to apply, he must give written notice to the contractor to that effect. It is possible that the employer might wish to reserve some decisions to himself or to specify that another person shall act for him in relation to particular clauses. For example, he might specify that a quantity surveyor acts for him in every clause which has a particular cost aspect. Clause 12 and clause 30 are obvious candidates for this treatment. There is no reason why article 3 should not be redrafted to allow the employer to appoint a number of persons as his agents for particular purposes as set out in the Employer's Requirements. To avoid confusion, it is essential that the respective powers and duties are clearly set down. For a comprehensive list of the employer's/the employer's agent's powers and duties see Figs 5.1 and 5.2.

5.2 Express and implied terms

Like the contractor, the employer is governed by express and implied terms. In most instances, an implied term will be excluded or modified by express terms of the contract. However, there are two terms of great importance concerning the employer which the law will imply into every building contract irrespective of its express terms. These terms are fundamental to the proper carrying out of the contract.

Fig. 5.1
Employer's powers.

Clause	*Power*	*Comment*
2.4.1	Agree contractor's proposed amendment or decide how discrepancy must be dealt with.	After contractor has notified employer of proposed amendment to deal with discrepancy within Employer's Requirements not dealt with by Contractor's Proposals.
2.4.2	Decide between discrepant items or accept the contractor's proposed amendment.	After contractor has notified employer of proposed amendment to remove discrepancy within Contractor's Proposals.
4.1	Issue instructions to the contractor on any matter in respect of which he is empowered to do so by the contract. Issue a written notice to the contractor requiring compliance with an instruction. Employ and pay others to give effect to the instruction. Recover all costs in connection therewith by deduction from monies due or to become due to the contractor under the contract or as a debt.	The contractor must immediately comply with such instructions *except* one requiring a change within clause 12.1.2, in which case he has the right of reasonable objection. If the contractor fails to comply with the seven day compliance notice.
4.2	Invoke the dispute resolution procedures.	If the employer wishes, a decision as to whether the specified clause provision in fact authorises the instruction. The request must be made before the contractor complies with the instruction.

Fig. 5.1 *Contd*

Clause	*Power*	*Comment*
4.3	Confirm in writing any non-written instruction which has been given. Confirm any oral instruction in writing at any time before the final account and final statement become conclusive as to the balance due under the contract.	Within seven days of its issue. This is a long-stop provision.
6.1	Consent to any amendment proposed by the contractor for removing the divergence and note the amendment on the specified documents.	Consent must not be unreasonably withheld.
8.1.1	Consent to the substitution by the contractor of anything in the Employer's Requirements, the Contractor's Proposals or the specifications referred to in clause 5.3.	Consent must be in writing and must not be unreasonably delayed or withheld. The consent does not relieve the contractor of his other obligations.
8.2	Request the contractor to provide vouchers to prove that materials and goods comply with clause 8.1.	
8.3	Issue instructions requiring the contractor to open up for inspection, testing, etc. of work, goods and materials.	The cost will be added to the contract sum unless the tests, etc. prove adverse to the contractor or where such costs have been provided for in the Employer's Requirements or Contractor's Proposals.
8.4.1	Issue instructions requiring removal from site of any work, materials or goods not in accordance with the contract.	
8.4.2	Issue reasonably necessary instructions re change following clause 8.4.1 instruction.	After consulting the contractor. No addition to the contract sum is to be made and no extension of time given.

Fig. 5.1 *Contd*

Clause	*Power*	*Comment*
8.4.3	Issue reasonable instructions to open up work or carry out tests to establish to employer's satisfaction the likelihood of further similar non-compliance.	After having due regard to the code of practice appended to the conditions.
8.5	Issue instructions re change reasonably necessary.	After failure to comply with clause 8.1.3. After consulting contractor no addition is to be made to the contract sum and no extension of time is to be given.
11	Have access to the work and to the workshops and other places of the contractor where work is being prepared for the contract.	This refers to the employer's agent and any other person authorised by the employer or the employer's agent. The right is exercisable at reasonable times, that is, during normal working hours.
12.2.1	Issue instructions effecting a change in the Employer's Requirements.	A change is a variation. No change may be effected (unless the contractor consents) which is or which will make necessary an alteration or modification in the design element of the works.
12.4A4.3	Refer the contractor's price statement or the amended price statement to the adjudicator under clause 39A.	To the extent that the amended price statement is not accepted by the contractor.
15	Consent in writing to the removal of unfixed materials and goods delivered to, placed on or next to the Works.	Consent must not be unreasonably withheld.
16.2	Issue instructions that defects, etc. be not made good.	An appropriate deduction must then be made from the contract sum.

Fig. 5.1 *Contd*

Clause	*Power*	*Comment*
16.3	Issue instructions that defects, etc. be made good.	No such instructions can be issued after delivery of the schedule of defects or after 14 days from the expiry of the defects liability period.
17.1	Take partial possession of the Works before practical completion.	With the contractor's consent.
18.1.1	Assign the contract.	Only if the contractor consents in writing.
18.1.2	Assign to any transferee or lessee the right to bring proceedings in the employer's name or enforce any contractual terms made for the employer's benefit. The assignee is stopped from disputing enforceable agreements reached between employer and contractor related to the contract.	If the employer transfers leasehold or freehold interest or grants a leasehold interest in the whole of the Works. If made prior to the date of the assignment.
18.2	Consent in writing to subletting all or part of the Works, including design.	Consent must not be unreasonably withheld or delayed.
21.1.2	Require the contractor to produce documentary evidence that clause 21.1.1.1 insurances are properly maintained.	The power must not be exercised unreasonably or vexatiously.
21.1.3	Effect the necessary insurance if the contractor has failed to insure or continue to insure.	The premiums may be deducted from monies due or to become due to the contractor or can be recovered as a debt.
21.2.2	Approve insurers	In regard to insurance against damage to property other than the Works - employer's liability.

Fig. 5.1 *Contd*

Clause	*Power*	*Comment*
21.2.3	Insure against damage to property other than the Works – employer's liability.	If the contractor fails to insure when so stated in the Employer's requirements.
22A.2	Approve the contractor's insurers and accept deposit of policies and premium receipts. Insure against all risks and deduct sums from monies due or recover them as a debt.	In regard to insurance against all risks to be taken out by the contractor. If the contractor has failed to insure or continue to insure.
22A.3.1	Inspect documentary evidence or the policy.	If the contractor maintains a policy independently of his obligations under the contract and it is in joint names.
22C.4.3.1	Determine the contractor's employment under the contract. Invoke the dispute resolution procedures.	If the Works are damaged by clause 22C.4 risks and it is just and equitable to do so. Within seven days of the contractor serving notice determining his employment.
22D.1	Require the contractor to accept the quotation in respect of liquidated damages insurance.	
22FC.3.2	Employ other persons to carry out remedial measures and deduct all costs from monies due to the contractor or recover them as a debt.	If the contractor fails to do so.
23.1.2	Defer giving possession for not more than six weeks.	Where clause 23.1.2 is stated in the appendix to apply.
23.2	Issue instructions regarding the postponement of any design or construction work to be executed under the contract.	
23.3.2	Use or occupy the site of the Works before the issue of practical completion certificate.	With contractor's written consent.

Fig. 5.1 *Contd*

Clause	*Power*	*Comment*
24.2.1	Give written notice requiring the contractor to pay liquidated damages *or* give written notice of intention to deduct liquidated damages.	The employer's notices under clause 24.1 and 24.2.1 must have been given.
25.3.2	Fix an earlier completion date than that previously fixed under clause 25.2 if it is fair and reasonable to do so in the light of any subsequently issued omission instructions.	This can only be done after an extension of time under clause 25.2.
25.3.3	Review extension of time granted and fix a new completion date or confirm existing. Fix an earlier date for completion.	May be carried out at any time after completion date if this occurs before practical completion, but no later than 12 weeks after practical completion. Having regard to instructions requiring an omission issued after the last extension of time.
27.2.1	Serve written notice on the contractor by actual, special or recorded delivery specifying the default.	If the contractor: • wholly or substantially suspends design or construction of the Works without reasonable cause; *or* • fails to proceed regularly and diligently; *or* • does not comply with a written notice to remove defective work; *or* • fails to comply with the assignment or subcontracting clause; *or* • fails to comply under the contract with the CDM regulations.

Fig. 5.1 *Contd*

Clause	*Power*	*Comment*
27.2.2	Determine the contractor's employment by written notice served by actual, special or recorded delivery.	The notice must not be given unreasonably or vexatiously. It can be served only if the contractor continues his default for 14 days after receipt of the default notice. The notice must be served within 10 days of the expiry of the 14 days.
27.2.3	Determine the contractor's employment by written notice served by actual, special or recorded delivery.	If the contractor ends the default or the employer does not determine and the contractor thereafter repeats the default.
27.3.3	Reinstate the contractor's employment in agreement with the contractor.	If the contractor is the subject of certain insolvency events which result in automatic determination.
27.3.4	Determine the contractor's employment.	Where the contractor's employment is not automatically determined after an insolvency event.
27.4	Determine the contractor's employment under this or any other contract.	Where the contractor is guilty of corrupt practices. No procedure is prescribed but determination should be effected by written notice served by recorded or special delivery as a precaution.
27.5.2	Make a 27.5.2.1 agreement with the contractor.	On the continuation or novation or conditional novation of this contract.
27.5.3	Make an interim arrangement for work to be carried out.	In the period before a 27.5.2.1 agreement is made.
27.5.4	Take reasonable measures to ensure that site materials, the site and the Works are adequately protected.	From the date when the employer may first determine the contractor's employment under clause 27.3.4.

Fig. 5.1 *Contd*

Clause	*Power*	*Comment*
	Deduct the reasonable cost of taking such measures from any monies due or to become due to the contractor or recover as a debt.	
27.6.2	Employ and pay others to carry out and complete the design and construction of the Works and make good defects and make use of the contractor's equipment, etc.	
27.6.3	Require the contractor to assign to him the benefit of any subcontracts, etc. Pay direct any supplier or subcontractor.	Unless determination was due to certain types of insolvency.
27.4	Require the contractor to remove from the Works his temporary buildings, etc. and in default to remove and sell the contractor's property.	
27.7.1	Decide not to have the Works carried out and completed.	The employer must notify the contractor within six months of determination.
28A.1	Give seven days written notice of determination.	If the carrying out of substantially the Works is suspended for a period named in appendix 1 by reason of: • *force majeure; or* • loss by specified perils; *or* • civil commotion; *or* • delay in receipt of development control permission; *or* • employer's instructions resulting from local authority or statutory undertaker's negligence; *or*

Fig. 5.1 *Contd*

Clause	*Power*	*Comment*
		• hostilities, *or* • terrorist activity.
29	Have work not forming part of the contract carried out by the employer or his licensees.	If the Employer's Requirements so provide and give the contractor the necessary information. If not, the contractor's consent is required under clause 29.2.
30.3.4	Give written notice to the contractor specifying amount proposed to be withheld or deducted from the due amount, the grounds and amount of withholding attributable to each ground.	Not later than five days before the final date for payment.
30.4.1	Deduct and retain the retention percentage, holding the same as trustee.	
30.5.6	Give written notice to the contractor three months after practical completion. Prepare or have prepared a final account and statement.	If the contractor does not submit final account and statement within three months of practical completion. If the contractor does not so submit.
S3.2	Consent to the appointment of manager. Consent to replacement of manager.	Must be in writing. For this clause to apply, the provisions must be stated in the Employer's Requirements. Must be in writing and not be unreasonably withheld or delayed.
S3.3	Request the manager and any of the contractor's servants, subcontractors, etc. to attend meetings.	Must be reasonable.
S3.4	Inspect the manager's records.	

Fig. 5.1 *Contd*

Clause	*Power*	*Comment*
S4.3.1	Consent to a person selected.	Following S4.2.2(b) change requiring the contractor to select another person to carry out named subcontract work.
S4.4.1	Consent to the determination of named subcontractor's employment.	
S6.5	Either: • instruct compliance and S6 shall not apply; *or* • withdraw the instruction.	If the parties cannot agree within 10 days of receipt of estimates following an instruction.
S7.4	Request such information as he may reasonably require in support of the contractor's estimate.	Within 21 days of receipt of an estimate of loss and/or expense under S7.2 or S7.3.
Fluctuation clauses are not covered in this table.		

Fig. 5.2
Employer's duties.

Clause	*Power*	*Comment*
1.4	Immediately notify the contractor in writing of the name and address of the new planning supervisor or the new principal contractor.	If he replaces the existing planning supervisor or the principal contractor.
2.3.1	Issue instructions on divergences between the Employer's Requirements and the definition of the site boundary. The instruction is deemed to be a change and is valued accordingly.	

Fig. 5.2 *Contd*

Clause	*Power*	*Comment*
2.3.2	Give written notice to the contractor specifying the divergence.	If such a divergence is found.
2.4.1	Agree the proposed amendment or decide how the discrepancy is to be dealt with.	Where there is a discrepancy in the Employer's Requirements not addressed by the Contractor's Proposals.
2.4.2	Decide between discrepant items or otherwise accept contractor's amendments as proposed.	Where there is a discrepancy in the Contractor's Proposals and the contractor has informed him of it. If such a discrepancy is found, the notice should be given immediately.
2.4.3	Give the contractor written notice specifying a discrepancy.	If found.
4.2	Specify in writing the contract clause empowering the issue of an instruction.	On the contractor's written request.
4.3	Issue all instructions in writing.	There is a procedure for the confirmation of oral instructions.
5.1	Be custodian of the Employer's Requirements, Contractor's Proposals and the contract sum analysis.	These documents must be made available for the contractor's inspection at all reasonable times.
5.2	Provide the contractor with a copy of the contract, the Employer's Requirements, Contractor's Proposals and Contract Sum Analysis certified on behalf of the employer.	This must be done immediately after the contract is executed unless the contractor has previously been provided with the documents.
5.6	Not to divulge to third parties or use any of the specified documents supplied by the contractor other than for the purposes of the contract.	

Fig. 5.2 *Contd*

Clause	*Power*	*Comment*
6.1.2	Immediately give written notice to the contractor of any divergence between statutory requirements and either the Employer's Requirements or the Contractor's Proposals.	The Employer's Requirements include any change (clause 12) and the duty arises if such a discrepancy is found.
6A.1	Ensure that the planning supervisor carries out all his duties under the CDM Regulations and that the principal contractor carries out all his duties under the CDM Regulations.	Where the contractor is not the planning supervisor. Where the contractor is not the principal contractor.
6A.3	Notify the planning supervisor of amendments to the health and safety plan.	If notified by the contractor.
7	Define the boundaries of the site.	
12.2.2	Vary the terms of the instruction to remove the contractor's objection.	If the contractor objects to an instruction under his obligations under the CDM Regulations.
12.3	Issue instructions to the contractor on the expenditure of provisional sums included in the Employer's Requirements.	
12.4.2,A2	Notify the contractor in writing either: • the price statement is accepted; *or* • that it or a part of it is not accepted.	Within 21 days of receipt.
12.4.2,A4	Include in notification to the contractor the reasons for not having accepted the price statement. The reasons must be set out in similar detail to that given by the contractor.	If the employer does not accept.

Fig. 5.2 *Contd*

Clause	*Power*	*Comment*
	Supply an amended price statement which is acceptable to the employer.	
12.4.2,A7	Notify the contractor: • either that the amount in lieu of loss and/or expense is accepted or is not accepted and clause 26 applies; *and* • either that the time adjustment is accepted or is not accepted and clause 25 applies.	If the contractor has attached his requirements to the price statement. The employer must act within 21 days.
15.2	List materials or goods or items prefabricated for inclusion in the Works.	The list is supplied to the contractor and annexed to the Employer's Requirements.
16.1	Give the contractor a written statement of practical completion.	When the Works have reached practical completion and the contractor has complied sufficiently with clause 6A.5.1. The statement must not be unreasonably withheld or delayed.
16.2	Deliver to the contractor a schedule of defects which appear within the defects liability period.	The defects specified must be due to the contractor's failure to comply with his contractual obligations *or* to frost occurring before practical completion. The schedule of defects must be issued not later than 14 days after the defects liability period expires.
16.4	Issue a notice that the contractor has made good all defects in the schedule.	Once the contractor has discharged his liability. The notice must not be unreasonably withheld or delayed.
22.3	Ensure that the joint names policies referred to in clauses 22B.1 or 22C.2 either:	If clause 22B or clause 22C applies.

Fig. 5.2 *Contd*

Clause	*Power*	*Comment*
	• provide for recognition of each subcontractor as an insured; *or* • include insurer's waiver of rights of subrogation.	
22A.4.4	Pay insurance monies received to the contractor.	By instalments under clause 30 alternative B even if alternative A is applicable to other payments.
22B.1	Maintain proper insurances against all risks. Produce receipts, etc. to the contractor at his request.	Where the employer has undertaken the risk in the case of new Works.
22C.1	Maintain adequate insurances against specified perils.	In the case of existing structures.
22C.2	Maintain insurance against all risks for work of alterations or extensions.	In joint names.
22C.3	Produce insurance receipts, etc.	If the contractor so requests.
22D.1	Either inform contractor that insurance is not required or instruct him to obtain quotations. Instruct contractor whether or not employer wishes quotation to be accepted.	If Employer's Requirements and appendix state liquidated damages insurance may be required.
22FC.2.1	Comply with the Joint Fire Code and ensure such compliance by his servants and agents, etc.	
22FC.4	Indemnify the contractor for consequences of breach of the code.	To the extent resulting from employer's breach.
23.1.1	Give possession of the site to the contractor on the date for possession.	

Fig. 5.2 *Contd*

Clause	*Power*	*Comment*
23.3.2	Notify insurers under clause 22A or 22B or 22C.2 to .4 and obtain confirmation that use or occupation will not prejudice insurance.	
24.1	Issue written notice to the contractor stating that he has failed to complete the construction of the Works by the completion date. Issue further notice as necessary.	 After a new completion date has been fixed after the issue of a clause 24.1 notice.
24.2.2	Pay or repay liquidated damages to the contractor.	Where a later completion date is fixed under clause 25.3.3.
25.3	Make in writing a fair and reasonable extension of time for completion by fixing a later date as the completion date and stating which of the relevant events have been taken into account and the extent to which regard has been given to any omission instruction issued since the fixing of the previous completion date. Notify contractor in writing.	It must become apparent that the progress of the Works is being or is likely to be delayed; *and* the contractor must give written notice of the cause of the delay and supply supporting particulars and estimate; *and* the reasons for the delay must fall within the list of relevant events. If reasonably practicable having regard to the sufficiency of the contractor's notice, etc., the extension must be granted not later than 12 weeks from receipt of particulars, etc. or not later than the current completion date if that is less than 12 weeks away. If it is not fair and reasonable to fix a later date. Notice must be given not later than 12 weeks from receipt or before completion date, whichever is the earlier.

Fig. 5.2 *Contd*

Clause	*Power*	*Comment*
25.3.3	Write to the contractor either: • fixing a later completion date than that previously fixed if it is fair and reasonable to do so having regard to the relevant events; *or* • fixing an earlier completion date, likewise in the light of any omission instructions issued subsequently; *or* • confirming the completion date previously fixed.	It is a duty to review the situation whether the relevant event has been notified or not. No decision under clause 25.3 can fix a date earlier than the date for completion stated in the contract.
26.1	Reimburse the contractor for any direct loss and/or expense caused by matters affecting regular progress of the Works.	If the contractor makes written application within a reasonable time of it becoming apparent, and the necessary procedural and other conditions of the clause have been satisfied. It is implied that an ascertainment will be made as under JCT 98 – which should be dealt with in the Employer's Requirements.
27.6.7	Pay to the contractor any amount due to him after completion of the Works by others, less expenses, damages, etc.	
27.7.1	Notify the contractor in writing within six months of the date of determination. Send a written statement of account to the contractor.	If the employer decides not to have the Works completed. If the employer so notifies.
28.3.2	Immediately inform the contractor in writing if he has made a composition or arrangement with creditors or, if a company, made a proposal for voluntary arrangement for composition of debts or scheme of arrangement.	

Fig. 5.2 *Contd*

Clause	*Power*	*Comment*
28.4.3	Pay the contractor the retention deducted prior to the determination.	Within 28 days of determination. Subject to employer's right of deduction accrued before determination.
28.4.4	Pay the contractor the amount properly due.	After taking amounts previously paid into account. Payment must be made within 28 days of submission.
28A.5	Pay the contractor half the retention deducted prior to determination.	Within 28 days of determination. Subject to employer's right of deduction accrued before determination.
28A.6	Prepare an account. Pay the contractor the amount properly due.	With reasonable dispatch if contractor has discharged his obligation to provide documents within two months. After taking amounts previously paid into account. Payment must be made within 28 days of submission.
30.3.3	Give written notice specifying the amount of proposed payment, the basis of calculation and to what it relates. Pay the amount proposed.	Not later than five days after receiving application. No later than the final date for payment subject to clause 30.3.4.
30.3.5	Pay the amount stated in the interim application.	If no written notice given under clauses 30.3.3 and/or 30.3.4.
30.3.7	Pay the contractor simple interest.	If the employer fails to pay the amount due by the final date for payment. Payment is treated as a debt and the rate is 5% above Bank of England Base Rate.

Fig. 5.2 *Contd*

Clause	*Power*	*Comment*
30.4.2.2	Place the retention in a separate designated banking account and inform the contractor in writing that the amount has been so placed.	If the contractor so requests. The retention is to be banked at the date of each interim payment. The employer gets the interest.
30.6.1	Give a written notice to the contractor specifying the amount of payment proposed in respect of any balance due to the contractor in the final statement or in the employer's final statement.	Not later than five days after the final statement becomes conclusive as to the balance due.
30.6.3	Pay the balance stated as due in the final statement.	If the employer does not give the notices in clauses 30.6.1 and/or 30.6.2.
30.6.4	Pay the contractor simple interest.	If the employer fails to pay the amount due by the final date for payment. Payment is treated as a debt and the rate is 5% above Bank of England Base Rate.
31.3.2	Notify the contractor in writing that he intends to make the statutory deduction from payments due under the contract and give reasons for his decision.	If the contractor fails to provide a tax certificate.
31.5	Promptly send to the Inland Revenue any voucher given to him by the contractor.	
31.6.1	Notify the contractor of statutory tax deductions which he proposes to make.	The Inland Revenue regulations should be referred to in this complicated procedure.
31.7	Correct any errors or omissions in calculations or statutory deductions by repayment to or deduction from the contractor as appropriate.	

Fig. 5.2 *Contd*

Clause	*Power*	*Comment*
34.2	Issue instructions on antiquities found.	If the contractor reports a find of antiquities, etc. under clause 34.1.
39A.4.1	Give notice to the other party of his intention. Refer the dispute to the adjudicator. Include particulars of the dispute and summary of contentions relied on, statement of relief sought and any other material he wishes the adjudicator to consider.	Where the employer requires a dispute to be referred to adjudication.
39A.5.7	Meet own costs of adjudication.	Except that the adjudicator may direct that a party bears the costs of tests or opening up if required.
39B.1.1	Serve on the contractor a notice of arbitration.	If the employer wants a dispute referred to arbitration.
S1.3.1	Set out the matters in dispute in a statement to the adjudicator.	Not later than 14 days after giving notice that a dispute has arisen.
S1.7	Bear the adjudicator's fee equally with the contractor.	
S4.2.2	Either: • remove the reason for inability; *or* • omit the named subcontract work from the Employer's Requirements and issue instructions re the execution of such work.	If the contractor is unable to enter into a subcontract with the named person for bona fide reason.
S5.2	Correct errors in description or quantity in bills of quantity.	The correction is to be treated as a change in the Employer's Requirements.

Fig. 5.2 *Contd*

Clause	*Power*	*Comment*
S5.4	Provide amplification of any bills of quantities included in the Employer's Requirements.	Where clause 38 applies.
S7.4	Give the contractor written notice either: • that he accepts the estimate; *or* • that he wishes to negotiate and in default of agreement to refer the issue to the adjudicator; *or* • that clause 26 shall apply.	Within 21 days from receipt of clause S7.2 or S7.3 estimate.

The two terms are complementary and they may be expressed in the context of this contract as follows:

- The employer and his agent will do all that is reasonably necessary to enable the contractor to carry out and complete the Works in accordance with the contract: *Luxor (Eastbourne) Ltd* v. *Cooper* (1941).
- Neither the employer nor his agent must hinder or prevent the contractor from carrying out and completing the Works in accordance with the contract: *Cory (William) & Sons* v. *City of London Corporation* (1951).

These terms are capable of very broad interpretation and often form the basis of substantial claims by the contractor. For example, on one hand the employer must be sure to give the contractor all necessary decisions in good time, and ensure the site is available and that there is a good access; on the other hand, he must not close the access, refuse to give decisions and stop the work. If the employer fails to comply with these two implied terms, the contractor's obligation to complete by the contract completion date is removed and his duty is simply to complete the Works in a reasonable time. The problems this would cause are largely avoided in standard building contracts by the inclusion of a clause allowing extension of the period for carrying out the work for most of such failures (see Chapter 8).

5.3 *Instructions*

5.3.1 Procedure

The issue of instructions is covered by clause 4 which closely resembles its counterpart in JCT 98. All instructions issued by the employer or by the employer's agent must be in writing. The contract does not prescribe any special form for the purpose. It is sufficient if the words are presented to the contractor in permanent visible form and provided it is clear that the words are instructing the contractor to do something. Most instructions will be in the form of a letter, but they can be written on a pad on site, on the back of an envelope or on the side of a brick. The more bizarre methods are not advocated. If the employer's agent is an architect, he may well use a standard form for issuing instructions, but care must be taken to strike out the words 'Architect's Instruction' and substitute 'Employer's Agent'. Whoever issues instructions will find it helpful to use standard forms for the purpose because they make the checking process much easier at the end of the project.

After stating that all instructions must be in writing, the contract proceeds at some length to set out the procedure if an instruction is issued 'otherwise than in writing', i.e. orally (clause 4.3.2). Oral instructions are of no immediate effect and if the contractor complies with such an instruction, he does so at his own risk. Depending on the content of the instruction, the contractor may be in breach of his obligation to carry out the work in accordance with the contract documents. For example, if the employer gives an oral instruction requiring the enlargement of a restaurant terrace to seat 100 rather than 50 people and the instruction is not confirmed but the contractor complies, it will result in work which is not in accordance with the contract. In theory, the employer can order rectification or removal under clause 8.4, but in *G. Bilton & Sons* v. *Mason* (1957) it was held that a contractor's compliance with unconfirmed architect's instructions was a good defence to a claim for breach of contract. Evidence would have to be brought that the oral instruction was given. Under clause 4.3, after receiving an oral instruction, the contractor has seven days in which to write to the employer to confirm it. The employer then has a further seven days in which to dissent. If the employer does not dissent, the instruction takes effect, not from the date it was issued, but from the expiry of the employer's seven day dissent period.

If the employer confirms an oral instruction within seven days, the instruction takes effect from the date of confirmation and the contractor's duty to confirm is removed. If neither contractor nor employer confirms an oral instruction in writing but the contractor has nevertheless complied, clause 4.3.2.2 provides that the employer may confirm it in writing at any time up to the date at which the final account and final statement become conclusive. Strangely, in this instance, the instruction is then deemed to have been issued on the date the oral instruction was given. This is intended as a safeguard for the contractor and to provide a mechanism whereby the contractor can secure payment if he complies with an oral instruction. A contractor who leaves the confirmation of an oral instruction to this late date, however, is asking for trouble. With the best will in the world, which may not be much in evidence, the memory of the employer's agent will dim and he may even be replaced with another. There is generally little excuse for oral instructions. They are a sign of laziness - probably on both sides. It is good practice for the contractor to keep a duplicate book on site for the benefit of the employer's agent. Oral instructions can be jotted down, signed and dated and there is no need for delay or complex confirmations. In these days of the fax machine, the days of the telephoned instruction should be at an end.

In theory, there should be very few instructions issued by or on behalf of the employer under this form of contract, because the issue of many instructions removes much of the risk which the contractor otherwise takes in respect of the date for completion and the price. The reality can be different, perhaps because the employer has not properly finalised his requirements in the contract documents or perhaps because he does not appreciate the crucial differences between this and other forms of procurement (see Chapter 1).

Clause 4.1 provides that the contractor must forthwith comply with all instructions issued by the employer. The meaning of 'forthwith' is that he must comply as soon as he reasonably can: *London Borough of Hillingdon* v. *Cutler* (1967). There are a number of important provisos:

- The employer must be expressly empowered under the contract to issue the instruction in question. A list of instructions empowered by the contract is given in Fig 5.3. Clause 4.2 gives the contractor power to request the employer to specify in writing the empowering clause in the contract. The employer must comply as soon as he reasonably can and if the contractor then complies with the instruction, it will be deemed empowered by that clause for all the purposes of the contract. The most important 'purpose' is probably the valuation of change instructions,

Fig. 5.3
Employer's instructions empowered.

Clause	*Instruction*
2.3.1	In regard to discrepancy between Employer's Requirements and definition of the site boundary.
4	Employer's instructions in general.
6.3.3	After amendment to Employer's Requirements to which clause 6.1.1.1 refers becomes necessary to comply with statutory requirements.
8.3	To open up for inspection or carry out testing.
8.4.1	Removal from site of work, materials or goods.
8.4.2	Reasonably necessary change after defective work.
8.4.3	Reasonably necessary to open up and establish likelihood of non-compliance.
8.5	Reasonably necessary change after failure to comply with clause 8.1.3.
12.2.1	Effecting a change in Employer's Requirements.
12.3	Expenditure of provisional sums included in Employer's Requirements.
16.2	Schedule of defects.
16.3	Requiring defects, etc. to be made good.
22D.1	To obtain and accept liquidated damages insurance.
23.2	Postponing work.
27.6.4	To remove temporary buildings, etc. from the Works after determination.
34.2	Regarding antiquities, including excavation, examination or removal by third parties.
S4.2.2(b)	Omitting named subcontract work and regarding the execution of that work.
S6.5.1	Compliance with instruction.

clause 12.5. The effect of this clause is that even if the employer is wrong in believing that his instruction is empowered under the specified clause, the contractor is entitled to whatever benefits would flow from a properly empowered instruction if he queries it before compliance. As an alternative, the contractor or indeed the employer may invoke the relevant procedures for the resolution of disputes to decide whether the instruction is empowered by the specified clause. The relevant procedure usually will be adjudication unless both parties wish the matter to be referred immediately to arbitration or dealt with in litigation (whichever method is included in the contract). It is thought that the contractor has the right to await the outcome before complying. Sensibly, he may decide to comply pending the result of the dispute procedure if he can get written agreement from the employer that such compliance is without prejudice to his rights and remedies following the outcome.

- If the instruction is for a change which makes it necessary to modify or alter the design of the Works, the contractor's consent is required (clause 12.2.1). This provision is often overlooked by employer and contractor alike. Its effect is twofold: it helps to impress on the employer that changes should be the exception rather than the rule, and it affords the contractor, as designer, the opportunity to resist a change which will result in serious amendment to his design. However, the contractor must not unreasonably delay or withhold his consent, for example to get even with an employer with whom he has had a difference on some other matter.
- The contractor need not comply with an instruction requiring a change under clause 12.1.2 (that is to say in respect of obligations as to access or use of parts of the site, limitations of working hours or space or the carrying out and completion of the work in any specific order) to the extent that he makes a reasonable written objection to the employer. There are two points to note. Provided a reasonable objection is lodged, the contractor need not comply and it is not for the employer to decide what is reasonable. If the employer disputes the objection, it is a matter for the dispute resolution procedure. Therefore, such an objection can have expensive results: for the employer in delay to the Works if he seeks dispute resolution under the contract and loses; and for the contractor in the costs of proceedings and delay if the employer wins. If the contractor objects, sweet reason dictates that the parties sit down and sort out the problem. The phrase 'to the extent' means that if the contractor objects to a part of the

instruction, he is entitled to withhold compliance only from that part; he must comply with the rest of the instruction.

If the contractor, without proper grounds under the contract, does not carry out the instruction, clause 4.1.2 gives the employer the right to employ other persons to do whatever is necessary to carry out the instructions. Before so doing, the employer must give the contractor written notice requiring compliance with the instruction within seven days. The employer's rights become operative if the contractor fails to comply. The employer may deduct from the contractor all costs he incurs as a result or he may recover them as a debt. This will include the money paid to the other persons and all other costs such as professional fees, if appropriate, and the cost of such things as scaffolding, cutting out and reinstatement. The wise employer will obtain alternative quotations for carrying out the work, unless time precludes it, so that he can prove if necessary that he has paid the lowest practicable price: *Fairclough Building Ltd* v. *Rhuddlan Borough Council* (1985). The contractor is entitled to see a breakdown of the price. The employer may only deduct the additional cost to him of the contractor's failure to comply excluding the amount he would have had to pay the contractor for doing the work anyway. The deduction, which is simpler than suing for a debt, can be made under clauses 30.3.4 or 30.6.2 (see Chapter 10).

5.3.2 Specific instructions

The individual instructions empowered by the contract are discussed here.

Clause 2.3.1 In regard to discrepancy between Employer's Requirements and definition of the site boundary

The employer has a duty under clause 7 to define the site boundaries. In practice this will probably be done by the employer's solicitor. In this respect it should be noted that boundaries are notoriously difficult to settle with accuracy, particularly where old property is concerned. Once the employer has defined the boundary, presumably on a site plan, the contractor is entitled to proceed on the basis of that definition and if he commits an act of trespass solely as a result of a mistake in the employer's definition, the contractor would be able to look to the employer for an indemnity. If the parties agree that the contractor will provide the site, clause 7 must be amended so that it is the responsibility of the contractor to define the boundary. The contract empowers the employer to issue an instruction to correct a discrepancy between the Employer's Requirements and the definition of site boundary which in turn entitles the contractor to a valuation. In the event that the contractor provides the site, this clause must also be amended.

Clause 8.3 To open up for inspection or carry out testing

This clause gives the employer power to instruct that work covered up is to be opened up for inspection, or to arrange for the testing of any materials whether or not they are already built into the Works. It is a valuable power, because its very existence can dissuade the contractor from attempting to incorporate work or materials which are not in accordance with the contract, and also because it enables the employer to check that work covered up is correct and the materials used are not cheap substitutes. The power is not without its drawbacks of course, and rightly so. If the workmanship or materials which are the subject of the check are found to be defective, the contractor is to stand the cost of putting matters right. If, however, everything is found to be in accordance with the contract, the employer must pay the cost of testing or opening up and making good. In addition, the contractor may have grounds for an extension of time and reimbursement of direct loss and/or expense under clauses 25.4.5.2 and 26.2.1 respectively.

Clause 8.4 Defective work, materials or goods

The employer has wide powers to issue instructions in respect of defects. Defective work or materials is work or materials which are not in accordance with the contract. The clause closely follows the equivalent clause in JCT 98. In the first edition we remarked that the employer, in addition to instructing the removal from site, had the power to instruct its rectification. This was a simple and obvious improvement to JCT 98. There are clearly many instances when rectification rather than removal from site is indicated. For some inexplicable reason, JCT have now deleted that power and employers faced with, say, defective painting will be obliged to instruct removal from site instead of the more sensible rectification. It is perhaps an oversight that in clause 12.1.1, there is still reference to 'rectification pursuant to clause 8.4'.

As an alternative or in addition to that power, the employer, by clause 8.4.2, may issue an instruction requiring a change. If the change is reasonably necessary as a result of the instruction to remove, the contractor is not entitled to any additional payment, extension of time or loss and/or expense. If the change is not entirely necessary but only partly so, and the employer is taking the opportunity to change or incorporate other requirements, the contractor is entitled to payment, extension of time and loss and/or expense in respect of the part of the change instruction which is not reasonably necessary. There is a stipulation that the instruction may be issued only after consultation with the contractor. The intention is probably to allow the parties to agree on the extent of the instruction which is necessary. In practice, although the employer must consult, i.e. seek advice or opinion, there is no requirement that the employer must take heed of such advice or opinion.

A valuable, but sometimes controversial, power is given to the employer by clause 8.4.3 which entitles him to issue instructions, under clause 8.3, to open up for inspection or to test other parts of the building or other materials. The idea is that he is entitled to establish to his reasonable satisfaction whether there are any similar cases where work or materials are not in accordance with the contract. Whether or not the opening up or testing shows that similar defects exist, the contractor is not entitled to payment for carrying out the instruction and reinstatement. This is notwithstanding the provisions of the opening up clause 8.3 and the loss and/or expense clause 26. The contractor is entitled to an extension of time unless the work or materials were found not to be in accordance with the contract.

Most importantly before issuing the instruction, the employer must have had regard to a set of criteria attached to the contract and dubbed the 'Code of Practice'. The intention of the Code is to assist in the fair and reasonable operation of clause 8.4.3. It provides that the employer and the contractor should try to agree the extent of the opening up or testing and the way it is to be accomplished and the employer is to consider 15 criteria. They cover the kind of factors which might well give the employer cause for concern, for example, the importance of demonstrating that the failure is a one-off occurrence, the degree of significance of the failure in the context of the building as a whole and the implications on safety of a similar failure elsewhere, the standard of the contractor's supervision and any proposals which the contractor may make. In addition, the employer is to consider, as item 15, 'any other relevant matters', a category which could hardly be broader in this context.

The instruction must be reasonable in all the circumstances. It appears, by use of the words 'to the extent', that if the instruction is not reasonable in the amount of opening up or testing required, the contractor is entitled to payment, extension of time and loss and/or expense for the part of the instruction which is not reasonable. What is or is not reasonable is a matter for the arbitrator. It is likely that the employer has a great deal of scope in issuing instructions under this clause. The proviso that he must 'have regard' to the Code of Practice would be satisfied if he simply read it before issuing the instruction. The effect of this provision, like its fellow in JCT 98, is very onerous so far as the contractor is concerned.

Clause 8.5 Workmanlike manner

Clause 8.1.3 requires that all work must be carried out in a proper and workmanlike manner and in accordance with the health and safety plan. If the contractor fails to comply with the first part of the clause, the employer, under clause 8.5, may issue an instruction, including requiring a change, if it is reasonably necessary as a consequence of the failure. This is very similar to the provisions of clause 8.4.2 following the discovery of work or materials not in accordance with the contract. In similar fashion, the contractor is not entitled to extension of time, loss and/or expense or any other payment. The only safeguard for the contractor is that the employer may not issue an instruction under this clause without first consulting him. That proviso is likely to be small comfort to him. When this clause was introduced, the Joint Contracts Tribunal indicated that it had taken account of the decision in *Greater Nottingham Co-operative Society Ltd* v. *Cementation Piling and Foundations Ltd and Others* (1988), which considered whether a nominated sub-contractor had liability to the employer in respect of bad workmanship. This clause should avoid the problem encountered in that case by making clear that, if the employer has to issue an instruction requiring a change on account of the contractor's failure to proceed in a workmanlike manner, the contractor can secure no advantage, financial or otherwise, as a result of such instruction.

Clauses 12.2 and 12.3 Effecting a change in the Employer's Requirements and expenditure of provisional sums

The employer is entitled to issue instructions to change his requirements under clause 12.2.1, but it is doubtful whether he is entitled to issue an instruction directly to vary the work or design. The consequence may be much the same, but not inevitably so. Clause 12.2.1 appears to leave the door ajar, but reference to 12.1, the definition of change, makes clear that only the requirements and not the design may be changed. For example, in the case of a hotel lobby, there may be lounge type seating for 20 persons requested in the Employer's Requirements and provided in the Contractor's Proposals. The employer may wish to increase the seating capacity to 30 people. Under a traditional contract, the employer would inform his architect who would redesign that portion of the building so as to accommodate the increased seating requirement. The revised drawing showing exactly what was to be done would be issued to the contractor together with an architect's instruction to carry out the revised work. Under WCD 98, however, the employer could simply issue a change instruction to the contractor, stating that he required an additional ten seating spaces. In that instance, it would be for the contractor to look at the implications for the design and, if appropriate, refuse to comply (see also section 5.3.1). He would only refuse, of course, if it was reasonable to do so.

It is difficult to state precisely what would be reasonable and each situation would be judged on its merits. In the example, the addition of a few seats seems hardly likely to provoke such a response from the contractor. It may be, however, that in order to accommodate the extra seats, the lobby would require enlarging, which might in turn create difficulties elsewhere. The contractor would probably be justified in withholding consent until such time as he could explain to the employer the full cost and other implications of the instruction, and possibly beyond that if the change involved a virtual redesign or massive rebuilding. If there was no problem, the contractor would simply take the instruction, carry out the redesign and, if so required in accordance with clause S2, submit for approval and then proceed. It appears that, if clause S2 is inapplicable (see section 2.8) the contractor has no obligation to obtain the employer's consent to the revised design, although it may be wise to do so. The employer can exercise greater or less control over the result of his instruction by varying the degree of detail he includes in the instruction requiring a change. He might instruct 'ten extra seats in the lobby' or 'ten extra seats in the lobby, four of which should be facing the reception area and situated along the northern wall'.

Clause 12.2.2 applies if the contractor is also the planning supervisor. It gives him the right to object to an instruction in accordance with his duties under regulation 14 of the CDM Regulations. He may only object under this clause if the instruction requires a change or if it concerns the expenditure of a provisional sum in the Employer's Requirements. The objection must be in writing. Once the objection is lodged, the employer must vary the terms of the instruction to remove the objection to the reasonable satisfaction of the contractor. Despite the provisions of clause 2.1, the contractor is not obliged to comply with the instruction until it has been so varied.

Clause 12.3 provides that the employer may issue instructions regarding the expenditure of provisional sums included in the Employer's Requirements. Note that he may not issue instructions in respect of any sums in the Contractor's Proposals and if the contractor has included such sums, they must be transferred to the Employer's Requirements before the contract is executed. Although the contract provides for provisional sums, it is in the employer's best interests to include as few such sums as possible. Every sum introduces an element of uncertainty in price and time which moves the risk towards the employer and away from the contractor roughly in proportion to the value of the provisional sum in relation to the contract sum.

Clauses 16.2 and 16.3 Schedule of defects and requiring defects to be made good

Under the provisions of these clauses, the employer is entitled to serve on the contractor, at the end of the defects liability period, a list of defects which have appeared during that period and also to issue such instructions as he may deem necessary for the correction of defects during the period. These clauses are dealt with in detail in Chapter 7.

Clause 22D.1 To obtain and accept liquidated damages insurance

The power under this clause is exercisable if the employer wishes to insure against the loss of liquidated damages following an insurance loss caused by a specified peril. The implications of this insurance are discussed in Chapter 11.

Clause 23.2 Postponing work

This clause provides a valuable power to the employer to postpone not only the carrying out of the work, but also the design of the work or any part. This reflects the contractor's responsibilities under this form of contract. This power must be exercised with caution. Postponement entitles the contractor to an extension of time under clause 25.4.5.1, direct loss and/or expense under 26.2.4

and to determine his employment under clause 28.2.2.2 if the postponement affects substantially the whole of the works for a period exceeding the period entered in the appendix. It is worth noting that a court or arbitrator may decide that a postponement instruction has been issued even though the employer has not used those words or referred to this clause. A court will look at whether a letter or instruction issued by the employer amounts to a postponement instruction, although issued for some other purpose: *Holland Hannen & Cubitts (Northern) Ltd* v. *Welsh Health Technical Services Organisation and Another* (1981) and *M. Harrison & Co (Leeds) Ltd* v. *Leeds City Council* (1980). Although not without doubt, it seems unlikely that a contractor can look to benefit from a postponement instruction which arises as a result of some defect in the work: *Gloucestershire County Council* v. *Richardson* (1967).

Clause 34.2 Regarding antiquities, including excavation, examination or removal by third parties

Under this clause, the employer has power to instruct the contractor to take specific action in respect of antiquities which he has reported under the provisions of clause 34.1. Notwithstanding the general nature of his power under this clause, it is specified that the instructions may require the contractor to permit the examination, excavation or removal of the object by a third party. Although the third party is deemed to be a person for whom the employer is responsible for the purposes of clause 20 in respect of injury to persons and property and indemnity to the employer, it is not envisaged that the third party will be employed by the employer under clause 29 and, therefore, provision for loss and expense and extension of time is made in clauses 34.3.1 and 25.4.5.1 respectively.

5.4 *Powers*

5.4.1 General

Under traditional forms of contract, the employer has few express rights of importance other than the obvious right to receive the building, completed in accordance with the contract on the due date. Under this form, however, in addition to the employer's powers to issue instructions, the contract confers some other substantial powers which are worthy of mention. A power exists whenever the contract states that the employer may do something. The employer's powers are summarised in Fig. 5.1.

5.4.2 Access

Clause 11 is one of only two clauses which expressly refer to the employer's agent. It provides that the employer's agent and any person authorised by the employer or his agent must have access to the Works, workshops or any other places where work is being prepared for the contract. The contractor is obliged to insert a term in appropriate subcontracts so as to obtain similar rights for the employer and his representatives. He must go further and do everything reasonably necessary in order to give effect to those rights. The employer may take advantage of his powers under this clause at all reasonable times. In this context a reasonable time would be during normal working hours. There is just one important proviso: the contractor and the subcontractor may impose whatever reasonable restrictions are necessary in order to protect their proprietary rights in the work to which the employer has access. This is a vital safeguard at a time when increasing amounts of building components are of a specialist nature and trade secrets must be safeguarded.

If the employer is to appoint a clerk of works in addition to the employer's agent, the clerk of works should be authorised to act under this clause as well as having his existence and duties delineated in the Employer's Requirements.

5.4.3 Partial possession

The employer's power to take part of the Works into possession before practical completion of the whole is governed by clause 17. Although many employers seem to assume that this power is unfettered and subject only to their wishes, such is not the case. The contractor has power to refuse consent. Consent must not be unreasonably withheld, but in practice it can be very difficult for the employer to maintain that the refusal is unreasonable in the face of the contractor's insistence that partial possession will hamper his progress. Of course, generally the contractor will be delighted to secure the advantages which flow from partial possession (see Chapter 7). This clause is not intended for use where the employer knows before he invites tenders that he wishes to take possession of the building in parts. The sectional completion supplement which properly identifies the appropriate parts should be used for that purpose. The partial possession provision is intended for use only where the employer decides during the progress of the Works that partial possession of a particular part is desired.

5.4.4 Effect insurance

The insurance provisions are noted in detail in Chapter 11. It should be noted that the employer has important powers to scrutinise insurance policies and to take out and maintain insurance if the contractor fails to do so. Under clause 22C.4.3.1, the employer may determine the contractor's employment if it is just and equitable following the discovery of loss or damage caused by any of the insured risks. For a fuller discussion of this provision see Chapter 12.

5.4.5 Deferment of possession

If the employer fails to give possession on the date specified in the contract, it is normally a serious breach of contract which the employer cannot overcome by making an extension of time in the absence of express provisions: *Freeman & Son* v. *Hensler* (1900). The consequences are damages for the breach and possibly repudiation if the failure is severe. Although so serious, there are numerous instances where the employer offends in this way. Sometimes, employer and contractor agree informally that possession may be late, but strictly a special agreement should be entered into. Clause 23.1.2, therefore, gives the employer an important power to defer the giving of possession for a period which must not exceed six weeks, but may be whatever lesser period is stated in the appendix. The employer must state in the appendix whether this provision is to apply. In view of the consequences of failure to give possession on the due date, it is considered vital that the appendix should state that this clause should apply unless the employer is absolutely certain that there can be no difficulties. It is thought unlikely that contractors increase their tenders significantly, if at all, to take account of the risk. Of course, deferment has its consequences too, but they are regulated by the contractual machinery and provided the deferment does not exceed the stated period, the contractor will be entitled to an extension of time under clause 25.4.14 and loss and/or expense under clause 26.1.

5.4.6 Deduction of liquidated damages

If the contractor fails to complete the Works by the date for completion stated in the appendix or by any extended date, the employer is entitled to recover liquidated damages at the rate stated in the appendix. This can have severe consequences for the contractor, especially where the amount of liquidated damages is fixed as a substantial sum per day or per week. The exercise of the power is circumscribed by two very important preconditions: the employer must have issued a notice under clause 24.1 that the contractor has failed to complete by the due date, and the employer must have given notice of his intention to recover liquidated damages. If a new date for completion is fixed, the non-completion notice is automatically cancelled and the employer must issue a new notice. Any amount of liquidated damages which has been recovered must be repaid, but there is no requirement for interest. If, however, the employer has recovered liquidated damages unlawfully, for example because the liquidated damages clause is defective or the notice of non-completion has not been issued, the contractor may have a claim for recovery of interest as special damages for the breach: *Department of the Environment for Northern Ireland* v. *Farrans (Construction) Ltd* (1982).

In *A. Bell and Son (Paddington) Ltd* v. *CBF Residential Care and Housing Association* (1989) the judge confirmed that both notices of non-completion and a written requirement for payment were preconditions. This clause has been substantially redrafted to comply with the Housing Grants, Construction and Regeneration Act 1996 Part II, and a full discussion is in section 8.4.2.

It is important to remember that the employer or the employer's agent, even if he is an architect, has no duty to act fairly between the parties in issuing the notice of non-completion on which the recovery of liquidated damages is based. Such a notice is not of binding effect until arbitration, as would be a similar notice by the architect under the provisions of JCT 98: *J.F. Finnegan Ltd* v. *Ford Seller Morris Developments Ltd* (1991). Thus, if disputed, it seems that liquidated damages could not be recovered from the retention fund until arbitrated upon.

5.4.7 Review extensions of time

Under the provisions of clause 25.3.3, the employer is empowered to review the extension of time situation and either confirm the date for completion previously fixed or fix a new date earlier or later than the previous date. This valuable power can be used by the employer to prevent time becoming at large. For a further discussion see Chapter 8.

5.4.8 Work not forming part of the contract

The contractor has the right to exclusive possession of the site while he is carrying out the building contract. This is subject to some exceptions (see Chapter 7). If it were not for the inclusion of clause 29 in the contract, the employer would have no right to have work carried out by other persons on the site until the contractor had finished. Apart from legalities, it makes sense that the contractor would not be able to get on and complete his work properly and within the contract period if constantly interrupted by other persons tramping across the area where he is working. Employers, however, frequently wish to engage others to do particular parts of the work and they want far more control over these persons than would be the case if they were simply named subcontractors under supplementary provision S4 (see Chapter 6).

Clause 29 gives the employer power to engage others for particular work subject to certain conditions. There are two situations envisaged by the contract:

- If the employer has included in his Requirements sufficient information about the particular work so as to enable the contractor to complete the contract Works in accordance with the contract, the contractor must allow the particular work to be carried out by others (clause 29.1);
- If the employer has made no reference to the particular work in his Requirements, but he wishes to have such particular work carried out by others, the employer may arrange for the carrying out of the work if the contractor consents. There is a stipulation that the contractor must not unreasonably withhold his consent (clause 29.2).

In a contract whose philosophy is to place as much responsibility as possible in the hands of the contractor, a clause like this seems rather out of place. Clearly, the contractor will be able to organise resources, hit targets and generally manage the project more effectively if he has complete control over the site and resources. Employers, therefore, would be advised to consider very carefully whether they really want to exercise the power contained in this clause which will almost certainly affect some element of the project, whether financial or concerning time or quality. If the employer feels it necessary to use this clause, there will usually be a price to pay. Clauses 25.4.8 and 26.2.3 entitle the contractor to an extension of time and loss and/or expense respectively in appropriate cases (see Chapter 8).

Clause 29 refers to 'work not forming part of the contract'. That is precisely correct. The work referred to in this clause is not work which the employer can require the contractor to carry out even, it is thought, by using the powers in clause 12.2. Neither can this clause be used to allow the employer to omit work from the contract under clause 12.2 and give it to others using clause 29. This is particularly so when the person concerned is not an independent professional, but the person directly interested: *Vonlynn Holdings Ltd* v. *Patrick Flaherty Contracts Ltd* (1988); *AMEC Building Contracts Ltd* v. *Cadmus Investments Co Ltd* (1997).

5.5 *Employer's duties*

5.5.1 General

Duties are normally indicated in the contract clauses by the use of the word 'shall'. If the employer fails to carry out any of his duties under the contract, it will be a breach of contract for which the contractor will have a remedy in damages (which may, of course, be nominal), quite apart from any specific remedy prescribed under the terms of the contract itself. Some of the employer's most important duties are discussed below. A full list of those duties is given in Fig. 5.2.

5.5.2 Notices

Since there is no independent architect administering the contract, it falls to the employer or his agent to issue any notices required under the contract. The most important notices relate to discrepancies which the employer may discover under clauses 2.3, 2.4.3 and 6.1.2, the statement of practical completion under clause 16.1, notices of making good defects under 16.4 and 17.1.2 and notice of failure to complete on the due date. A full list of the notices and statements to be issued by the employer is given in Fig. 5.4. The effect of such

Fig. 5.4
Statements and notices to be issued by the employer.

Clause	*Statement or notice*
2.3.2	Notice of divergence.
2.4.1	Notice of agreement or decision.
2.4.3	Notice of discrepancy.
4.1.2	Notice requiring compliance with an instruction.
6.1.2	Notice specifying divergence between statutory requirements and Employer's Requirements or Contractor's Proposals.
12.4.2,A2	Notice that price statement is or is not accepted.
12.4.2,A7.1	Notice regarding acceptance of contractor's extension of time and loss and/or expense requirements.
16.1	Statement of practical completion.
16.4	Notice of making good of defects.
17.1.2	Notice of making good of defects.
22C.4.3.1	Notice of determination.
24.1	Notice of failure to complete on due date. Further notice of failure to complete on due date.
24.2.1	Notice that payment, etc. may be required. Notice requiring payment or deducting.
27.2.1	Notice of default.
27.2.2	Notice of determination.
27.3.4	Notice of determination after insolvency.
27.7.1	Notice that Works are not to be completed.
28A.1.1	Notice giving seven days to determination.
30.3.3	Notice specifying the amount to be paid.
30.3.4	Notice of intention to withhold payment.
30.5.6	Notice if contractor does not submit the final account and final statement.

Fig. 5.4 *Contd*

Clause	*Statement or notice*
30.6.1	Notice specifying the amount to be paid.
30.6.2	Notice of intention to withhold payment.
39A.4.1	Notice of intention to refer to adjudication.
39B.1.1	Notice of arbitration.
S7.4	Accepting estimate; *or* wishing to negotiate or in default referring to dispute resolution procedures or applying clause 26; *or* applying clause 26.

notices is discussed in detail in the appropriate chapters, but it should be appreciated that if the employer neglects to issue a notice at the right time, the consequences will always be serious.

5.5.3 Possession

The employer must give possession of the site to the contractor on the due date unless the deferment clause 23.1.2 applies and it has been properly operated by the employer.

5.5.4 Extensions of time

The employer has a duty to make extensions of time in accordance with the procedures set out in clause 25 (see Chapter 8) This task would normally be undertaken by the architect if a traditional contract was used. Since the employer or his agent can in no sense be considered to be disinterested, it is suggested that the fixing of a new date for completion under this form of contract will be examined by the courts or an arbitrator with correspondingly more care than would be the case under a traditional form.

5.5.5 Payment

Perhaps the most important duty of the employer is to pay the contractor in accordance with the terms of the contract (see Chapter

10). Failure to pay is dealt with in clause 28.2.1.1 where the contractor has power to determine his employment under certain conditions. The contractor has no right at common law to stop work just because he has not been paid what he considers to be the correct amount: *Lubenham Fidelities & Investment Co* v. *South Pembrokeshire District Council and Wigley Fox Partnership* (1986). This can be a very serious matter for the contractor who may not be able to fund continuation of the project in the face of the employer's breach. In *Lubbenham*, the contractor was unable to determine under the contract provisions because the employer had correctly paid the amount shown on the architect's certificate. It was the amount which was clearly and demonstrably wrong. There are no certificates under WCD 98 and the contractor is potentially in a very strong position. It should also be noted that the courts do appreciate the contractor's problems where payment is withheld and such withholding may be held to amount to a repudiatory breach if so repeated that the contractor has no realistic expectation that he will ever be paid: *D.R. Bradley (Cable Jointing) Ltd* v. *Jefco Mechanical Services* (1988). However, under the provisions of section 112 of the Housing Grants, Construction and Regeneration Act 1996, the contractor has the right to suspend performance in such circumstances on seven days notice (clause 30.3.8).

Payment by cheque is probably good payment although in theory the payment is not made until the cheque is cleared through the bank. Contractors faced with an employer who simply does not pay are in serious difficulties. An employer can often resist payment if he can show that he has a strong bona fide case for doing so: *C.M. Pillings & Co Ltd* v. *Kent Investments Ltd* (1985); *R.M. Douglas Construction Ltd* v. *Bass Leisure Ltd* (1991). However, the excellent and rapid process under the adjudication procedure in clause 39A could ensure rapid attention to the problem (see Chapter 13).

If the contractor opts not to seek adjudication or to suspend the work, the contractor's best way forward is to operate the determination provisions, provided he is sure that the failure can properly be brought under this clause. On receipt of the default notice, the employer may immediately pay. If the employer does not pay within 14 days, the contractor can determine his employment and he may be able to negotiate suitable terms for continuance of work thereafter. If terms are not agreed, the employer must settle the account and if he fails to do so, referral to adjudication or arbitration as appropriate can proceed without the contractor having the burden of carrying out the work.

CHAPTER SIX

SUBCONTRACTORS AND STATUTORY REQUIREMENTS

6.1 *General*

In this chapter we consider the subject matter of clauses 6 (Statutory requirements), 18 (Assignment and subcontracts) and 29 (Execution of work not forming part of the contract). There is no provision for nomination under this contract, but the supplementary provisions allow for 'persons named as sub-contractors in Employer's Requirements'. How far such provisions permit the employer to impose his own choice of subcontractor on the main contractor will be considered later in this chapter.

Subcontracting, assignment and novation are often confused. Before considering the contract provisions in detail, it is important to understand the difference between these terms. Conveniently, they were set out with admirable clarity by Lord Justice Staughton in *St Martins Property Corporation Ltd and St Martins Property Investments Ltd* v. *Sir Robert McAlpine & Sons Ltd and Linden Gardens Trust Ltd* v. *Lenesta Sludge Disposals Ltd, McLaughlin & Harvey PLC and Ashwell Construction Company Ltd* (1992) in the Court of Appeal:

'(a) Novation This is the process by which a contract between A and B is transformed into a contract between A and C. It can only be achieved by agreement between all three of them, A, B and C. Unless there is such an agreement, and therefore a novation, neither A nor B can rid himself of any obligation which he owes to the other under the contract. This is commonly expressed in the proposition that the burden of the contract cannot be assigned, unilaterally. If A is entitled to look to B for payment under the contract, he cannot be compelled to look to C instead, unless there is a novation. Otherwise B remains liable, even if he has assigned his rights under the contract to C...

(b) Assignment This consists in the transfer from B to C of the benefit of one or more obligations that A owes to B. These may be obligations to pay money, or to perform other contractual promises, or to pay damages for a breach of contract, subject of course to the common law prohibition on the assignment of a bare course of action. But the nature and content of the obligation, as I have said, may not be changed by an assignment. It is this concept which lies, in my view, behind the doctrine that personal contracts are not assignable... Thus if A agrees to serve B as chauffeur, gardener or valet, his obligation cannot by an assignment make him liable to serve C, who may have different tastes in cars, or plants, or the care of his clothes...

(c) Sub-contracting I turn now to the topic of sub-contracting, or what has been called in this and other cases vicarious performance. In many types of contract it is immaterial whether a party performs his obligations personally, or by somebody else. Thus a contract to sell soya beans, by shipping them from a United States port and tendering the bill of lading to the buyer, can be and frequently is performed by the seller tendering a bill of lading for soya beans that somebody else has shipped.'

6.2 *Subcontractors*

6.2.1 Assignment

Clause 18.1.1 contains the usual restriction on the assignment of the contract by either party without the written consent of the other. In the *St Martins* case, the House of Lords (1993) held that this clause effectively prevents the benefit of the contract being assigned. For example, the employer might wish to sell the building before the final account and final statement become conclusive or the contractor may wish to assign the right to receive payment in return for a cash advance. With clause 18.1.1 in place, consent would have to be given by the other party in each case, but see clause 18.1.2 considered below. There is nothing to prevent a party refusing consent on grounds which might be considered unreasonable. This can pose real problems and if the employer might possibly wish to assign the benefit of the contract (i.e. sell or otherwise transfer the property to another) before practical completion, an amendment to the clause is indicated.

The general law forbids assignment of the burden of a contract, so this clause is superfluous in that regard. For the contractor to effectively transfer to another the duty to carry out the work set out in the contract documents, or for the employer to transfer the duty to pay for such work, would require the consent of both parties to the contract together with the consent of the party who is to shoulder the burden in place of either contractor or employer.

Clause 18.1.2 will only apply if so stated in appendix 1. It contemplates the situation where the employer sells the freehold or the leasehold interest in the premises comprising the Works to a third party, or where he grants a leasehold interest in the premises. In any of these instances, the employer may assign to that third party the right to bring proceedings in the employer's name and to enforce any of the terms of the contract. There is a proviso that the third party cannot dispute any agreement which is legally enforceable and which is entered into between the employer and the contractor before the assignment. This clause does not give the employer the right to sell the premises before he has received them from the contractor at practical completion, therefore it does not conflict with clause 18.1.1. However, once the employer has received the building and disposed of it by sale or lease, it enables the purchaser to act as if he was the employer so far as the benefits of the contract are concerned. For example, the obligation to pay the contractor remains with the employer, but the purchaser can enforce the defects liability provisions. However, if the employer and the contractor have entered into an agreement under which the contractor is not obliged to make good certain defects and no monetary deduction is to be made, it is binding on the purchaser of the premises under clause 18.1.2.

6.2.2 Subletting

An important difference between this contract and JCT 98 is that clause 18.2.3 allows the contractor to sublet design provided he has the written consent of the employer. In contrast to assignment, the employer may not unreasonably delay or withhold consent. If the employer does give his consent, the contractor's obligations under clause 2.5 are not affected (see section 3.2). In practice, the employer will normally give his consent readily, because few contractors keep high calibre design teams on the staff and an employer may have selected a particular contractor partly on account of the prestigious design team he has assembled. If that is the situation, the contractor would be well advised to have a suitable amendment made to the printed form to allow such subletting without the requirement for consent.

Professional fee agreements can be ill-suited to the design and build concept. Even such things as reference to the 'client' have to be understood in a new light as meaning the contractor. Therefore, references to the 'contractor' become confusing. Many contractors who carry out design and build contracts regularly have their own standard forms of agreement for the design team. Those who do not have such forms should seriously consider acquiring them.

Clause 18.2.1 is a prohibition on the subletting of the whole or any part of the Works without the written permission of the employer. Once again, such consent must not be unreasonably delayed or withheld. There is no requirement that the contractor must inform the employer of the names of subcontractors. It is merely consent to the fact of subcontracting which is required. It would be reasonable for the employer to refuse to give consent until the name and perhaps other details of the prospective subcontractor were made known. To avoid dispute, the employer could state in his Requirements that such information has to be submitted before consent will be given. There is a proviso that the subcontract must provide for the employment of any subcontractor to determine immediately on the determination of the contractor's employment under the main contract for any reason. This is to prevent a difficult situation arising in which the contractor's employment is determined under the main contract, as a result of which the subcontractor can bring an action against the contractor for breach of the subcontract. The current position is that a contract cannot bind anyone who is not a party to it; this is the doctrine of privity of contract. Therefore, this provision will not bite on the subcontractor unless it is put in the subcontract by the contractor. This is the reason for clause 18.3.1. The easiest way to accomplish this is for the contractor to enter into each subcontract (other than with the design team) on the basis of DOM/2 published by the CC.

Clause 18.2.2 was introduced by Amendment 6 which states that the contractor remains wholly responsible for the carrying out and completion of the Works in accordance with clause 2.1 even though some or even the whole of the Works is sublet. This clause was inserted in most JCT contracts in an excess of caution following the decision in *Scott Lithgow* v. *Secretary of State for Defence* (1989). Since the employer has no contractual relationship with the subcontractor, he must look to the contractor if there is any defect in the subcontractor's work. This is so even if the subcontractor has gone into liquidation. The contractor will still remain responsible for his work.

It is known as the 'contractual chain'. At times such a chain can be a long one such as when sub-sub- or even sub-sub-subcontractors are involved. Responsibility for defects goes right up through the chain and redress goes down through the chain. If the chain is broken by insolvency, responsibility rests with the party on the side of the insolvency nearest the employer (see Fig. 6.1). If the contractor has become insolvent when a defect is found in a subcontractor's work, the employer has no contractual remedy and, following *Murphy* v. *Brentwood District Council* (1990), little hope of a tortious remedy. To overcome such problems, the employer may make the provision of an acceptable warranty on the part of the subcontractor a pre-condition to the giving of any consent to subletting. Alternatively, the employer may insert such a stipulation, accompanied by an example of the warranty required, in his Requirements.

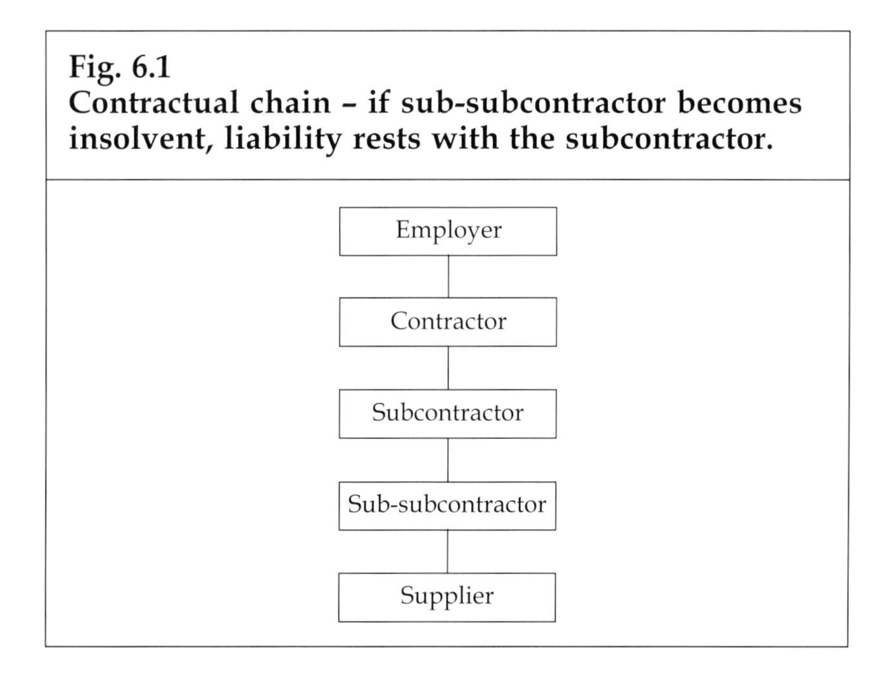

Fig. 6.1
Contractual chain – if sub-subcontractor becomes insolvent, liability rests with the subcontractor.

Clause 18.3 sets out certain provisions as a condition to sub-letting. Clause 18.3.1 states that each subcontract must provide that the employment of the subcontractor determines immediately determination of the contractor's employment takes place. Clause 18.3.2 states that the subcontract must include certain provisions regarding unfixed materials and goods delivered to the Works or adjacent to the Works. There are four terms:

(1) Such materials and goods must not be removed without the employer's consent unless for use on the Works
(2) Ownership of such materials and goods is to be automatically transferred to the employer after the value has been included in an interim payment
(3) If the contractor pays for such materials and goods before being himself paid by the employer, ownership passes to the contractor
(4) The operation of this clause is not to affect ownership in off-site materials passing to the contractor as provided in clause 15.2 of WCD 98.

The employer should ensure that these provisions are included in subcontracts and perhaps refuse to consent to subcontracting unless evidence of such inclusion is produced. Although standard subcontract DOM/2 contains such provisions, many contractors habitually subcontract using their own terms which not only do not contain such provisions, but also do not create a satisfactory 'back to back' subcontractual arrangement. Even if such provisions are included, they are ineffective to safeguard the employer from the perils of retention of title, if the subcontractor has bought the goods himself on terms that the supplier retains ownership until payment is made. Building contract chains are so long that it is virtually impossible to check down to the ultimate supplier that ownership has passed unimpeded up to the contractor.

Breach of the provisions of clauses 18.1.1 or 18.2.1 or 18.2.3 is a sufficient ground for determination by the employer under clause 27.2.1.4 (see Chapter 12).

6.2.3 Persons named as subcontractors in Employer's Requirements (S4)

This provision gives the employer some assurance that he can have certain parts of the project carried out by a subcontractor of his own choice. It has similarities to the naming provisions in IFC 98 and in ACA 2. The provision is only to apply if the Employer's Requirements state that certain work is to be carried out by a named person employed as a subcontractor by the contractor. The work is termed 'named sub-contract work' and the subcontractor is termed a 'named subcontractor'.

Clause S4.2.1 stipulates that the contractor must enter into a subcontract with the named person as soon as reasonably practicable after entering into the main contract. That is to say that the contractor must enter into the subcontract as soon as he can in practice, or allowing for the current situation. As soon as he has entered into the contract, the contractor will have a hundred and one things to do immediately; things which cannot be done before the main contract is executed. This subcontract will be only one of those things. It is always in the contractor's interest to execute subcontracts promptly so as to safeguard the price. Contractors often have a very loose arrangement with prospective subcontractors even though the tender price may be based on figures provided by them. For example, there is nothing in law to prevent a subcontractor from withdrawing his price before acceptance, even if he has stated in his quotation that it will remain open for a specific period (unless the contractor has paid to keep the option open). If this happened in the case of a named person, the result could be disastrous. At best, the contractor could be forced into the position of taking a loss on the subcontract. The contractor must notify the employer of the date of the subcontract. There is no requirement that this notification should be in writing, but it is sensible for the contractor to serve all notices in writing.

The contractor may not be able to enter into the subcontract, perhaps because the named person has withdrawn his quotation and refuses to submit another price. Whatever the reason, the contractor must notify the employer immediately to allow as much time as possible for action to mitigate the effect of the problem. The contractor must give the reason for his inability to conclude a subcontract. If the reason is bona fide, the employer may take alternative courses of action:

- If the reason is connected to the item in the Employer's Requirements, the employer may issue a change instruction to remove the reason; *or* he may issue a change instruction to omit the named subcontract work from his Requirements.
- If the employer chooses the second option above he may issue a further instruction requiring the contractor to select another person to carry out the work subject only to the employer's reasonable approval of that person. There seems no reason in principle why the contractor should not choose himself in an appropriate case.
- He may state that the work is to be carried out by a directly employed person as referred to in clause 29 (see section 6.5).

The employer may not name a substitute person in the change instruction.

Whether or not the reason advanced by the contractor for inability to enter into the subcontract is bona fide need not be left to the opinion of the employer or to his agent. If a dispute develops over the point, it may be referred to adjudication under the provisions of clause 39A.

The subcontract determination situation is dealt with in clause S4.4. The contractor may not determine the named subcontractor's employment for default unless he first obtains the employer's consent which must not be unreasonably withheld or delayed (clause S4.4.1). If the contractor proceeds to carry out the determination, he is responsible for completing whatever work is left unfinished. Although the employer is not to issue a change instruction to cover the situation, the work required to complete is to be treated as though it resulted from a change instruction. There are two exceptions:

- If the determination is the result of the contractor's default, i.e. if the subcontractor determines; *or*
- If the contractor has not obtained the employer's consent under clause S4.4.1.

Clause S4.4.3 is obscurely drafted, but it seems that the contractor must pay the employer any amounts which he has recovered, or which he could have recovered from the defaulting subcontractor, using reasonable diligence. What qualifies as reasonable diligence will vary with the circumstances. The amounts to be recovered are those which are legally due to the contractor as a result of the determination to reduce the cost of completion. This will only apply if the determination is due to the subcontractor's default. To avoid the subcontractor successfully contending in any proceedings that the contractor has suffered no loss and, therefore, that he has nothing to recover, the contractor is obliged by clause S4.4.4 to insert an appropriate clause in the subcontract. The clause must state that the subcontractor, having notice of clause S4, undertakes not to contend that the contractor has suffered no loss and that his liability should be reduced or extinguished in any way. It is thought that such a clause would be effective in practice: *Haviland and Others* v. *Long and Another, Dunn Trust Ltd* (1952).

6.3 Statutory requirements

Statutory requirements are dealt with by clause 6. It is a difficult clause, because although it is modelled on the equivalent clause in

JCT 98, there are important additions and changes which reflect the particular philosophy of this contract. The contractor's principal obligation is contained in clause 6.1.1.2. This clause will apply except to the extent that the Employer's Requirements specifically state that they comply with statutory requirements (clause 6.1.1.1). For example, if the Employer's Requirements say that the particular requirements as stated for an auditorium comply with statutory requirements, the contractor is entitled to assume that they do so comply and he may complete the design on that basis. If there is later found to be a failure to comply, the employer and not the contractor will be responsible for the cost of correcting the error. The contractor is not entitled to assume compliance for any part of the building other than the specific part stated in the Requirements – in this case, the auditorium. It should be noted that this sets out the position between the employer and the contractor, but the contractor cannot shelter behind this proviso so far as the statutory authorities are concerned. It would seem, however, that the provision would be sufficient, in most instances, to enable the contractor to recover from the employer any costs incurred in making the work covered by clause 6.1.1.1 comply with statutory requirements.

The contractor must comply with any Act of Parliament, instrument, rule or order or any regulation or byelaw of a local authority or statutory undertaker who has jurisdiction with respect to the Works or with whose systems the Works will be connected (water, electricity and drainage systems are obvious examples). That is said to include development control requirements and the contract refers to all these regulatory matters as statutory requirements. Development control requirements are defined in clause 1.3 as 'any statutory provisions and any decision of a relevant authority thereunder which control the right to develop the site'. They are often referred to for ease as 'planning requirements' and in a narrow sense that is correct. However, there are other statutory provisions which control the right to develop the site in a particular way. Examples of such provisions are the Fire Regulations and the requirement for an entertainment licence for an appropriate development. Thus, for a particular development, the failure to satisfy any statutory provisions which determine whether a site can be developed for a particular purpose would amount to a failure to satisfy development control requirements.

It is clear that the contractor's obligation is to comply with statutory requirements, not only in the design he is to complete, but also in the whole of the construction including construction of part of the design which may already be in the Employer's Requirements. The contractor must give the employer all statutory approvals received. He must also submit all notices in connection with statutory requirements.

Other than where the Employer's Requirements specifically state otherwise, it is clear that the contractor bears responsibility for compliance. This is emphasised by clause 6.1.2 which explains the procedure if either the employer or the contractor finds a divergence between statutory requirements and either the Employer's Requirements (which includes any changes) or the Contractor's Proposals. Whoever finds the divergence must immediately notify the other in writing, stating the divergence. Then the contractor must write to the employer with his proposals for dealing with the divergence at his own cost. The employer may not unreasonably delay or withhold his consent and the contractor must complete both the design and construction according to the amendment. Provided that the amendment deals with the divergence, it seems that the employer will have no grounds to withhold consent. If the employer dislikes the proposed amendment, he must issue a change instruction and pay the cost. The contractor is not required to pay the cost if the situation is covered by clause 6.3 (see below).

There is a very curious provision: the employer must note the amendment on 'the documents referred to in clause 5.1'. The documents referred to are the contract documents. Therefore, the provision allows the employer to amend the contract documents to correct a divergence after the contract has been executed. It seems that the intention is to prevent the contractor being able to claim payment later on the ground that the work represented in the amendment is not included in the contract. It not only appears to us to be unnecessary, it can be dangerous. It is not clear why the employer should be able to amend the contract documents on account of a divergence caused by a change instruction. This provision could bear some adjustment. Although under clause 12.1.1, a change may mean a change in the employer's requirements which necessitates an alteration in the design or quality or quantity of the Works, the alteration to the design in response to the instruction is entirely a matter for the contractor. He is entitled to be paid for correctly responding to the instruction. Therefore, if there is a divergence between the design response to a change instruction and statutory requirements, the responsibility must lie with the contractor.

Clause 6.3 is in three parts and it appears to modify the contractor's obligations under certain circumstances, although the drafting is not such as to encourage understanding by the people who will be using it.

Clause 6.3.1

If after the base date there is a change in statutory requirements (which include development control requirements) which makes it necessary to amend the Contractor's Proposals, the amendment is treated as a change instruction under clause 12.2. This means that the contractor will be entitled to payment for the change.

Clause 6.3.2

If after the base date there is a decision made by the appropriate authority for development control requirements and the decision sets out terms of a permission or approval which make it necessary to amend the Contractor's Proposals, the amendment is treated as a change instruction under clause 12.2. This is subject to anything that the employer may have stated to the contrary in his Requirements. This last provision is intended to enable the employer to place the risk of satisfying development control requirements on the shoulders of the contractor. On its own, it is doubtful that it is enough to carry out that purpose.

Clause 6.3.3

If it becomes necessary to amend that part or whole of the Employer's Requirements which the employer has stated specifically complies with statutory requirements, the employer must issue a change instruction for the purpose.

The intention behind the drafting of these clauses appears to be that if after the base date there is a change in statutory requirements, the contractor is entitled to be paid the cost of dealing with the change. For example, there could be an amendment to the Health and Safety at Work Act or to the building regulations or to one of the Planning Acts which affects the building for which the contractor has already entered into a contract. The contractor will be obliged to comply with the change in statutory requirements, but he is paid as though the employer had issued an instruction requiring that change. The same approach is applied to a permission or approval given after base date. The most common situation will be a decision given about reserved matters in a planning approval. The local authority may attach conditions and the contractor has to make amendments to his proposals to satisfy them. Once again, the amendment is treated as though the employer gave an express instruction for it and the contractor is entitled to be paid accordingly. There is provision for the employer to make the contractor take that risk by an appropriate statement in the Employer's Requirements. This is unlikely to be popular with contractors, because it is asking the contractor to budget for the unknown. There is a great difference between statutory requirements such as Building Regulations and development control requirements such as planning requirements. Whether or not a building will satisfy the Building Regulations is a matter which the contractor should be able to determine when he designs. He may commence building after simply serving the appropriate notice on the local authority. It is not possible to say whether a building will be given planning permission either in whole or in respect of any part and until permission is obtained, the contractor may not commence building. The last part of clause 6.3.2 may be contrary to the Unfair Contract Terms Act 1977 as an exclusion of liability which may not satisfy the test of reasonableness.

The contractor is entitled to an extension of time if there is any delay in him receiving necessary permission or approval from a statutory body (clause 25.4.7). The contractor must have taken all practicable steps to avoid or reduce the delay. In practice that means that the contractor must have applied at the right time and not left it until there was little chance of getting the permission in time and the delay must not be simply because the contractor defaulted in some way that an ordinary competent builder would have avoided. He may also have an extension of time under clause 25.4.13 following the situations in clauses 6.3.1 and 6.3.2 discussed above. Again, there is the proviso that the contractor must have taken all practicable steps to avoid or reduce the delay.

It is clear that the contractor is not to suffer a time penalty for changes in statutory requirements or delays in receiving permissions even, it seems, where the Employer's Requirements preclude payment for complying with terms in permissions relating to development control requirements under clause 6.3.2. The difference between predictable statutory requirements and unpredictable development control requirements is highlighted in clause 26, where delays in approvals for the latter, but not the former, is ground for an application for direct loss and/or expense if incurred (clause 26.2.2). Although the draughtsman's intention may have been to separate the risks in this way, it appears that the contractor has good grounds for making application if a clause 6.3.1 situation arises, because clauses 6.3.1 and 6.3.2 stipulate that the amendments are to be treated 'as if it were an instruction of the Employer under clause 12.2'. Clause 26.2.6 expressly makes provision for employer's instructions under clause 12.2 to be grounds for direct loss and expense applications. The complexities do not end there. If delay in the receipt of any permission or approval for development control requirement purposes results in the whole or substantially the whole of the Works being delayed for a period noted in appendix 1, the employer or the contractor may determine the contractor's employment under the contract (clause 28A.1.1.4).

The statutory requirement implications in any project entered into using this contract form deserve very careful consideration. This particularly applies to development control requirements as defined in clause 1.3. The employer who does not ensure that all such requirements are satisfied before executing the contract is taking a great risk. The employer cannot remedy the problem by simply deleting the extension of time, loss and/or expense and determination subclauses at the same time as inserting a statement in his Requirements that the contractor must satisfy development control requirements. If he followed that course of action, a failure to obtain planning permission would render the contract frustrated. That is the very reason why the determination provisions allow for determination if permission is unduly delayed and it is noteworthy that, in such circumstances, the contractor is denied the right to claim loss and/or expense arising out of the determination.

If it becomes necessary for the contractor to comply with statutory requirements as an emergency, he must supply the minimum work and materials necessary to ensure compliance (clause 6.1.3). The contractor must immediately inform the employer of the emergency and the steps he has taken to deal with it. Unlike the situation under JCT 98, the contractor is not entitled to payment for compliance and, therefore, no mention is made of it. However, if the emergency compliance related to a matter which the employer had expressly stated complied with statutory requirements under clause 6.1.1.1, the contractor would be entitled to payment and the employer must issue an instruction under clause 6.3.3.

Clause 6.2 provides that the contractor must pay all fees and charges in connection with statutory requirements and the contractor must include for them in his price unless they are stated as a provisional sum in the Employer's Requirements. The contractor's obligation is not simply to pay the fees and charges, but to indemnify the employer against all liability for them. Therefore, if the contractor fails to pay a charge and the employer is obliged to pay it together with a penalty, the contractor must reimburse the employer for both charge and penalty.

6.4 *The Construction (Design and Management) Regulations 1994*

Breach of the Construction (Design and Management) Regulations 1994 (1995 in Northern Ireland) will be a criminal offence, but except for two instances, a breach will not give rise to civil liability. Thus one person cannot normally sue another for breach of the Regulations. Compliance with the Regulations is made a contractual duty so that breach of the Regulations is also a breach of contract. This is likely to cause problems for the employer.

Article 7.1 has been introduced. It assumes that the contractor will be the planning supervisor. If one looks carefully, it is possible to see the word 'or', enabling the user to insert an alternative name. Article 7.2 defines the principal contractor as the contractor. It is not thought that these articles are sufficient in themselves to bind the contractor to the employer for the purpose of carrying out these functions and a separate contract for these services should be executed. They are certainly not sufficient if a third party is engaged as planning supervisor. It also follows that determination of the contractor's employment under this contract will not automatically determine his engagement as either planning supervisor or principal contractor and express terms must be written into the ancillary contracts to achieve automatic termination.

There are sundry definitions and words which make clear that the Works must be carried out in accordance with the health and safety plan. Grounds for determination (failure to comply with the Regulations) are included in the list in both employer and contractor determination clauses (clauses 27 and 28).

Clause 6A.1 provides that the employer 'shall ensure' that, where the planning supervisor and principal contractor are not the contractor, they will carry out their duties in accordance with the Regulations. There are also provisions that the contractor, if he is the planning supervisor and/or the principal contractor, will comply with the Regulations (clause 6A.2). The contractor must also ensure that any subcontractor provides necessary information. Compliance or non-compliance by the employer with clause 6A.1 is a 'relevant event' and a 'matter' under clauses 25 and 26 respectively. Lest the significance is missed, what this means is that the employer must ensure that, where the planning supervisor and/or the principal contractor is not the contractor they perform correctly and if they do not, or even if they do, any resultant delay or disruption will give entitlement to extension of time and loss and/or expense. This may well be a most fruitful source of claims for contractors. Every change instruction potentially carries a health and safety implication which should be addressed under the Regulations. The Regulations impose a formidable list of duties on the planning supervisor. Most of them are to be found in Regulations 14 and 15. Some of these duties must be carried out before work is commenced on site which may well present difficulties where the contractor is to be the planning supervisor and the contractor is not appointed until comparatively late in the process. It may be necessary to appoint a planning supervisor and then replace him with the contractor on acceptance of tender. If necessary actions delay the issue of a change instruction or once issued delay its execution, the contractor may be able to claim.

There may be rare occasions when the Regulations do not fully apply to the Works as described in the contract. If the situation changes due to the issue of an instruction or some other cause, the employer may be faced with substantial delay as appointments of planning supervisor and principal contractor are made and appropriate duties are carried out under the full Regulations.

Practice Note 27 has been published. It is very helpful, but it should be read with care and the knowledge that it has no legal weight. The way in which these clauses will work in practice will become clear in due course. It is certain that the key factor will be for employers, planning supervisors and principal contractors to structure their administrative procedures very carefully if they are to avoid becoming in breach of their contractual obligations.

6.5 *Work not forming part of the contract*

The employer has the right to make contracts with persons other than the contractor to carry out work on the site. The employer is not entitled to deduct work from the contractor so as to give it to another contractor. That would be a breach of contract which is certainly not contemplated by either clause 12.1.1.1 or clause 29: *Vonlynn Holdings Ltd* v. *Patrick Flaherty Contracts Ltd* (1988); *AMEC Building Contracts Ltd* v. *Cadmus Investment Co Ltd* (1997). Clause 29 provides for two situations:

- The Employer's Requirements may provide the contractor with very full information so that he can properly carry out the work required of him under the contract. They may also note that specific work is not to form part of the contract and will be carried out by others. In such an instance, the contractor must permit the specific work.
- The Employer's Requirements may not provide the full information noted above. In that case, the employer may still employ other persons to do work not included in the contract provided that the contractor gives his consent. The contractor must not unreasonably withhold his consent, but in the context of a design and build contract, it is suggested that it will be more difficult to prove that the contractor is unreasonable than under a traditional procurement system where responsibilities are already split.

Delays caused by persons engaged by the employer under clause 29 may give rise to an extension of time for the contractor under clause 25.4.8 or the payment of direct loss and/or expense under clause 26.2.3. The delays may be caused by persons failing to carry out the work, carrying it out slowly or simply carrying it out properly. For the employer to engage other contractors is ill advised at the best of times, but when a design and build contract is involved, it is like signing a blank cheque. After the employer has deliberately chosen a contract which puts as much responsibility as possible on the contractor's shoulders, it seems perverse to use this clause to take some of that responsibility away.

Persons engaged by the employer under this clause are deemed to be persons for which the employer is responsible for the purposes of the indemnity clause (clause 20) and not a subcontractor. That seems to be stating the obvious 'for the avoidance of doubt' as the lawyers say (see Chapter 11).

CHAPTER SEVEN
POSSESSION, PRACTICAL COMPLETION AND DEFECTS LIABILITY

7.1 *Possession and deferment*

Clause 23.1.1 provides that on the date for possession stated in the appendix, possession of the site must be given to the contractor. Possession is the next best thing to ownership. A person who is in possession of something, be it a car, a television set or a building site, has a better claim to it than any other person with the exception of the actual owner. The effect of this clause is to give the contractor a licence to occupy the site for the purpose of carrying out the construction. The owner has no general power to revoke the contractor's licence, but it would be brought to an end by completion of the Works or by determination of the contractor's employment under the provisions of the contract (see Chapter 12) or if the contractor's employment or the contract itself is lawfully brought to an end by some other circumstance. If there was not an express term giving possession, a term would be implied that the contractor must have possession in sufficient time to allow him to complete by the contract completion date: *Freemen & Son* v. *Hensler* (1900).

The contractor must have possession of the whole site: *Whittal Builders* v. *Chester-Le-Street District Council* (1987). Since the contractor is in control of the site, he may exclude all other persons from the site except those persons to whom the contract expressly allows access under clause 11 (see section 5.4.2) or those bodies which have powers of entry under statute. If the employer fails to give possession on the due date it is a breach of contract of a fundamental nature which entitles the contractor to damages and, because there is no provision for extension of time in such circumstances, time becomes at large. The contractor's obligation is simply to complete the Works within a reasonable time: *Rapid Building Group* v. *Ealing Family Housing Association* (1985). If the failure is prolonged, the contractor has the right to treat it as a repudiation of the contract on the part of the employer. To overcome this problem, WCD 98, in common with most JCT contracts, has a clause (23.1.2) which allows the employer to defer giving possession for a period which must not exceed six weeks. This clause applies only if so stated in the appendix and, therefore, it is vital that it is stated to apply or the employer will face serious consequences for any failure. A lesser period than six weeks may be specified in the appendix and, clearly, the shorter the period, the less the contractor will feel inclined to increase his tender figure. It is possible, of course, to amend the provision so that a very much longer period is specified, but the employer will be appropriately penalised in the contractor's tender. In practice, failure to give possession is either a matter of a few days or many weeks. Either the date is just missed, probably through some minor carelessness, or there is a major problem. The precise period for insertion is something to be discussed between the employer and his professional advisors.

Even where the employer has wisely stated that clause 23.1.2 is to apply, he will not escape some consequences of deferring possession. The contractor will be entitled to recover any direct loss and/ or expense he has incurred and he will be entitled to an extension of time. From the employer's point of view, the ability to make an extension of time for deferment of possession prevents time becoming at large and, therefore, preserves the employer's right to deduct liquidated damages for any culpable delay on the part of the contractor. There is no prescribed form for the notice of deferment and the employer need not give any reason although it would be courteous to do so.

Clause 23.3.1 provides that 'for the purposes of the Works insurances', the contractor retains possession of the site and the Works up to the date of the statement by the employer under clause 16.1, setting out the date of practical completion. Until then, the employer is not entitled to take possession of any part of the Works unless clause 17 (partial possession) has been operated. The drafting of this clause has been criticised because it appears to imply that the contractor retains possession merely for the purposes of the Works insurances. While the clause, when read alone, can be construed in this way, read in context with the rest of the contract, it is simply stating the position with the Works insurance in mind. The contractor retains possession for the principal purpose of carrying out the Works; the Works insurance is but a part of that purpose. It is the old argument that kettles may be blue, but not all blue objects are kettles.

Having established that the contractor retains possession until practical completion, clause 23.3.2 goes on to state that if the contractor gives his written consent, the employer may use or occupy the site or the Works or any part before practical completion. The employer may use or occupy for the purpose of storage of goods or otherwise – which is fairly broadly drafted. The reason for the clause appears to be to enable the employer to store goods on the Works without the necessity for operating the partial possession clause (clause 17). There is a procedure which must be observed before the contractor may give consent. Either the contractor or the employer must notify the insurers under the appropriate insurance provision (22A, 22B or 22C.2 to .4). If the insurers confirm that the use or occupation will not prejudice the insurance, the contractor may not withhold permission without good reason. In practice, finding a reasonable ground for withholding consent should not present too much of a problem.

Clause 23.3.3 stipulates that if the insurers have made a condition that an additional premium is required under clauses 22A or 22B, the contractor must notify the employer. The employer, in turn, must state whether he still requires use or occupation under this clause and if so, the additional premium must be added to the contract sum and if the employer requires a receipt, the contractor must provide one.

7.2 *Progress*

Clause 23.1.1 stipulates that the contractor must regularly and diligently proceed with the construction of the Works and complete them 'on or before' completion date. The precise meaning of 'regularly and diligently' has been the subject of some discussion. In *London Borough of Hounslow* v. *Twickenham Garden Developments Ltd* (1970), the judge considered the meaning for some time and concluded: 'At present, all that I can say is that I remain somewhat uncertain as to the concept enshrined in these words'. Perhaps more helpful are the observations of the Court in *Greater London Council* v. *Cleveland Bridge & Engineering Co Ltd* (1986) which were approved by the Court of Appeal. The Court was considering the meaning of 'due diligence and expedition' which was not actually included as an obligation of the contractor, but only in the negative way that failure to work with due diligence was a ground for determination:

> '...I would have held without hesitation, that due diligence and expedition must be interpreted in the light of the other obligations as to time in the contract. That seems to me to follow from

> the construction of the contract as a whole and from the considerations I have already mentioned . . . If there had been a term as to due diligence, I consider that it would have been, when spelt out in full, an obligation on the contractors to execute the works with such diligence and expedition as were reasonably required in order to meet the key dates and completion date in the contract.'

This has to be considered together with the general principle that a contractor is entitled to plan and carry out the work to suit himself provided that there is no provision to the contrary and that he meets the completion date: *Wells* v. *Army and Navy Co-operative Society* (1902). The employer may impose restrictions in his Requirements or in a change instruction under clause 12.1.2. The principle to be derived from those cases seems to be that it is difficult to successfully contend that a contractor is failing to proceed regularly and diligently provided he is doing some work on the site and provided that he can meet the completion date. However, the Court of Appeal in *West Faulkner* v. *London Borough of Newham* (1995) have helpfully defined 'regularly and diligently' in terms which make it easier for the employer to allege such failure:

> 'What particularly is supplied by the word "regularly" is not least a requirement to attend for work on a regular daily basis with sufficient in the way of men, materials and plant to have the physical capacity to progress the works substantially in accordance with the contractual obligations.
>
> What in particular the word "diligently" contributes to the concept is the need to apply that physical capacity industriously and efficiently towards the same end.
>
> Taken together the obligation upon the contractor is essentially to proceed continuously, industriously and efficiently with appropriate physical resources so as to progress the works steadily towards completion substantially in accordance with the contractual requirements as to time, sequence and quality of work.'

Where the sectional completion supplement is used with this form of contract, it indicates certain key dates which must be met. Provided such key dates are in the contract, they will be binding on both employer and contractor and the contractor's progress can be measured accordingly.

Under normal circumstances, the contractor's programme is not binding on either employer or contractor. It is very unusual for it to be made a contract document. It is quite clear, from clause 23.1.1, that the contractor must finish by the completion date, but he may finish before such date. The point was emphasised so far as the JCT Standard Form 1963 was concerned in *Glenlion Construction Ltd* v. *The Guinness Trust* (1987). By analogy, it is thought that the other holding in that case, that the architect is not obliged to produce information to suit the contractor's shortened work period, applies to decisions and approvals which the employer may have to give under this form.

The contractor's right to complete before the completion date can be a source of embarrassment to an employer who has scheduled his finances to cope with fairly equal amounts to be paid throughout the contract period. If he depends on income to fund payments, he may be driven into overdraft. It is difficult to avoid this situation, but it can be alleviated by the simple expedient of amending clause 23.1.1 to omit the words 'or before' after 'the same on', so that the contractor's obligation becomes to complete 'on the Completion Date'. If this course of action is taken, the contractor's attention should be particularly drawn to the change: *J. Spurling Ltd* v. *Bradshaw* (1956); *Interfoto Picture Library* v. *Stiletto Visual Programmes* (1988). It is, of course, still open to the contractor to complete most of the work and leave very little to finish in the last part of the contract period. To properly regulate the contractor's progress is possible, but it would require much more severe re-drafting.

Under the provisions of clause 23.2, the employer has the power to postpone any design or construction work which the contractor is to carry out under the terms of the contract (see section 5.3.2).

7.3 Practical completion

Practical completion is dealt with in clause 16.1. There is a significant difference between practical completion in this contract and in JCT 98. In JCT 98, a certificate is to be issued by the architect when, in his opinion, practical completion has been achieved. WCD 98 merely provides that the employer must give the contractor a written statement when the Works have reached practical completion and when the contractor has complied with clause 6A.5.1 or complied sufficiently with clause 6A.5.2 – in effect when the contractor has provided the employer with the health and safety file required by the CDM Regulations. Under JCT 98 it is a matter for the architect's opinion, while under WCD 98 it is a matter of fact.

Since the employer has merely to state fact, rather than certify opinion as an independent professional, the statement is not thought to be binding until arbitration: *J.F. Finnegan Ltd* v. *Ford Seller Morris Developments Ltd* (1991). The contract, however, states that for all purposes of the contract, practical completion is deemed to have taken place on the date named in the statement.

The introduction of a requirement to comply with clause 6A.5.1 and 6A.5.2 has introduced some ambiguity into this clause. What it amounts to is that for practical completion to be 'deemed' to have taken place for all the purposes of the contract, two criteria must be satisfied:

- The Works must have reached practical completion; *and*
- The contractor must have complied with clause 6A.5.1 and complied sufficiently with clause 6A.5.2.

One of these criteria is clearly practical completion of the physical Works. Just as clearly, one criterion is not practical completion. Yet the clause makes clear that practical completion will not be deemed to have taken place until both are satisfied. The inclusion of a 'deeming' provision has the effect that although both parties recognise that practical completion has not taken place on that day (indeed, it may have taken place in a physical sense some time earlier), they both behave as though it took place on the date stated: *Re Coslett Contractors Ltd* (1997). This may have repercussions on the liquidated damages position (see section 8.4).

The contract does not define what is meant by practical completion of the Works and there is conflicting case law. A sensible approach seems to be the one taken in *H.W. Neville (Sunblest) Ltd* v. *Wm Press & Sons Ltd* (1981) where it was held that at practical completion there should be no defects apparent although there might be trifling items outstanding. The idea was explained further in *Emson Eastern Ltd (In Receivership)* v. *E.M.E. Developments Ltd* (1991):

> 'I think that the most important background fact which I should keep in mind is that building construction is not like the manufacture of goods in a factory. The size of the project, site conditions, use of many materials and employment of many types of operatives makes it virtually impossible to achieve the same degree of perfection as can a manufacturer. It must be a rare new building in which every screw and every brush of paint is absolutely correct.'

A practical test was suggested by Lord Justice Salmon in the Court of Appeal considering the 1963 standard form of contract, and it was not disapproved on appeal to the House of Lords, in *Westminster Corporation* v. *J. Jarvis & Sons* (1969):

> 'The obligation upon the contractors under clause 21 to complete the works by the date fixed for completion must, in my view, be an obligation to complete the works in the sense in which the words "practically completed" and "practical completion" are used in clauses 15 and 16 of the contract. I take these words to mean completion for all practical purposes, that is to say, for the purpose of allowing the employer to take possession of the works and use them as intended. If completion in clause 21 meant completion down to the last detail, however trivial and unimportant, then clause 22 would be a penalty clause and as such unenforceable.'

If the contractor maintains that practical completion has taken place and the employer declines to give a statement to that effect, the contractor may refer the question to adjudication under clause 39A. In practice, the parties normally carry out a joint inspection. The contractor will be anxious to secure the statement that the Works have reached practical completion, because some very important consequences flow from it:

- The contractor's liability for insurance under clause 22A.1 ends
- Liability for liquidated damages under clause 24.2 ends
- Liability for frost damage ends (clauses 16.2 and 16.3)
- The employer's right to deduct full retention ends. Half the retention becomes due for release (clause 30.4)
- The three months period begins during which the contractor must submit the final account and final statement (clause 30.5.1)
- The period for final review of extension of time begins (clause 25.3.3)
- The defects liability period begins (clause 16.2).

7.4 Partial possession

The contract makes provision for partial possession by the employer in clause 17. It is not intended for use as a means of achieving sectional completion, because the contractor cannot be compelled to complete in predetermined parts; a sectional completion supplement is provided for that purpose. Clause 17.1 provides that the employer may take possession of any part of the Works before practical completion if the contractor's consent has been obtained. This power is said to be in spite of anything which may be expressed or implied elsewhere in the contract. Therefore, a conflict is avoided between this clause and, for example, clause 23.1.1 which effectively gives complete possession to the contractor until completion. The contractor's consent must not be withheld unreasonably. In contrast to the provisions for practical completion, it is the contractor who must 'thereupon issue to the Employer a written statement'. The statement must identify the part or parts of the Works taken into possession and the date possession was taken. The clause proceeds to refer to the matters so identified as the 'relevant part' and the 'relevant date' respectively. The contractor is not expressly required to estimate the value of the relevant part, but the making of an estimate is implied in the remainder of the clause.

Certain consequences follow after the contractor issues the written statement. Each of them is beneficial to the contractor and it is seldom that he will refuse to allow partial possession. They broadly echo the consequences of practical completion. Indeed, it is almost as if practical completion has been reached as regards the relevant part. The consequences are:

- The defects liability period commences and half the retention sum is released in respect of the relevant part (clause 17.1.1).
- After defects, notified under clause 16.2 or clause 16.3 in respect of the relevant part, have been made good, the employer must issue a notice to that effect (clause 17.1.2).
- From the relevant date, the obligation to insure the part taken into possession, whether it be by the contractor under clause 22A or by the employer under clause 22B or 22C.2 ends. If the employer is insuring under clause 22C, he must include the relevant part in his insurance of clause 22C.1, existing structures (clause 17.1.3).
- If the contractor becomes liable to pay liquidated damages after the relevant date, the amount of damages payable is to be reduced pro rata to the value of the relevant part as a proportion of the contract sum. This relatively simple concept is still expressed in a clause of awesome complexity despite much criticism (clause 17.1.4).

There is a school of thought which contends that the operation of this provision may be sufficient to change liquidated damages into a penalty on the basis that a genuine pre-estimate of loss in respect of a building as a whole may not be a genuine pre-estimate of loss when reduced simply on a pro rata basis. A relatively insignificant part of the building in terms of straightforward construction cost may be the most valuable in terms of use by the employer. Once that part is taken into possession, the reduction in liquidated damages does not properly reflect reality. Against such arguments may be set the thought that liquidated damages are commonly rather less than a genuine pre-estimate of loss.

7.5 *Defects liability period*

Respective liabilities during and after the defects liability period are much misunderstood. The period is to be named in appendix 1. It is important that a period is inserted, because failure to name a period will mean that the period will be six months from the date of practical completion. That may be perfectly satisfactory, but it is becoming common for employers to require 12 months as a more appropriate period, because it exposes the building to the full range of seasonal differences. Any defect in the Works is a breach of contract on the part of the contractor. The idea of the defects liability period is to allow a reasonable period for defects to become apparent and for the contractor to rectify them. This saves the employer the time and effort of rectification by another and taking legal action for the cost. From the contractor's point of view, he can rectify his own defects more cheaply than another contractor who is a stranger to the Works. If it were not for this clause, the contractor would have no right or duty to enter the site after practical completion. His licence would have expired (see clause 23.1 and section 7.1).

The employer has 14 days after the end of the defects liability period in which to deliver a schedule of defects to the contractor and the contractor has a 'reasonable time after receipt' in which to make good the defects. What is a reasonable time will depend on the number and the type of defects.

The contractor's liability for defects does not end when the defects liability period ends. What ends is simply his right to rectify the defects. It will be seen that even this right is severely circumscribed. The defects liability period is commonly referred to as the 'maintenance period'. Some other forms of contract, such as ACA 2 or GC/Works/1 edition 3, adopt that terminology. However, it is best avoided, since it implies the considerably more onerous duty to keep in pristine condition even though the contracts where it is used restrict the effect to the equivalent of defects liability in JCT contracts.

The defects which the contractor is to make good are spelled out in detail. They are 'Any defects, shrinkages or other faults'. Certain criteria must be satisfied. The defects must appear within the defects liability period and they must be due to the contractor's failure to comply with his obligations under the contract or due to frost which occurred before practical completion. The phrase 'defects, shrinkages and other faults', which might be taken to be extremely wide, is somewhat less so and 'other faults' is to be interpreted *ejusdem generis*. Thus, the faults must be of the same class as defects and shrinkages. The contractor's failure to comply with his obligations under this clause may be referable to design or construction. A defect which stems from some other cause, such as some inadequacy in the Employer's Requirements, is not covered by this clause. The contractor's design obligation under this contract should eliminate many of the arguments about defects. For example, it is not open to the contractor to argue that timber shrinkage is due to inadequate specification if the specification is part of his Proposals. The contractor's liability to make good frost damage is limited to damage caused by frost when the building was in his control. After practical completion, measures to prevent frost damage are the responsibility of the employer. In practice, it is usually easy to see the difference between frost damage before and after practical completion.

A strict reading of the clause suggests that defects which are outstanding at practical completion may not be considered as falling within this clause. This reinforces the definition of practical completion as being the point at which there are no apparent defects. However, to avoid absurdity it should be interpreted as including any defects which are apparent at practical completion: *William Tomkinson & Sons Ltd* v. *The Parochial Church Council of St Michael and Others* (1990). The employer must specify the defects which satisfy the criteria set out in the clause and he must deliver them to the contractor as an instruction no later than 14 days after the end of the period. The contractor's obligation is then to make good the defects specified within a reasonable time at his own cost. What is reasonable will depend on the number and kind of defects in the instruction. There is an important proviso that the employer may instruct that some or all of the defects are not to be made good and an appropriate deduction is to be made from the contract sum.

There is no definition of an 'appropriate deduction'. It is often contended by the employer that it is the cost to the employer of having the defects made good by others. The contractor, understandably, will argue that the deduction should be the cost which the contractor would have expended on making good. It seems unlikely that the employer's contention is correct, as a comparison with clause 4.1.2 demonstrates. Clause 4.1.2 deals with the situation where the employer has power to deduct all costs in connection with the engagement of others following the contractor's default in complying with an instruction in clear terms. Had such terms been intended to apply to a clause 16.2 situation, it would have been a simple matter to have repeated the provision. Probably the clue lies in the fact that if the employer instructs the contractor not to make good certain defects, it is not or not necessarily because the contractor has defaulted, but because the employer has chosen to exercise his right under the contract and in any event, the employer has an ordinary duty to mitigate his loss. This may be because the employer has lost all faith in the contractor's ability to rectify the defects, but it may be because the employer is prepared to put up with various minor imperfections to avoid disturbance to his office, factory or home. The appropriate deduction is thought to be the cost which the contractor would have incurred in rectifying the defect. We are strengthened in this view by the observations of Mr Justice Stannard in the *Tomkinson* case noted above when considering a very similar point:

> '... but the true measure of damages which governs this aspect of the case is not the church's outlay in remedying the damage, but the cost which the contractors would have incurred in remedying it if they had been required to do so.'

Important powers are conferred on the employer by clause 16.3. He may issue instructions for the making good of any defect which satisfies the same criteria as laid down in clause 16.2. These instructions may be issued 'whenever he considers it necessary to do so'. It is clear that this power is quite separate from the requirement to prepare a list of defects at the end of the defects period. Within a reasonable time, the contractor shall comply at no cost to the employer unless the employer instructs otherwise when an appropriate deduction from the contract sum must be made. The equivalent clause in JCT 98 contains a proviso that no such instruction may be issued after the architect (in that case) has delivered a schedule of defects 14 days after the end of the defects liability period. Clause 16.3 formerly had no such proviso and it followed that the employer could issue such instructions after delivering the schedule. A similar prohibition to the one contained in JCT 98 has now been inserted. Therefore, no such instructions can be issued after the earliest of either the delivery of the defects schedule or the expiry of 14 days following the end of the defects liability period.

When the contractor has made good all the defects notified to him by the employer under clauses 16.2 and 16.3, the employer must issue a notice to that effect. It must not be unreasonably delayed or withheld, because it affects the final account and final statement (see section 10.6). The contract states, unnecessarily, that making good defects is to be deemed to have taken place on the date named in the notice 'for all the purposes of this Contract' (clause 16.4).

Defects which appear after the end of the defects liability period are still the liability of the contractor. Although he can no longer demand the opportunity to make good as a contractual right, the principle of mitigation of loss will often mean that an employer will invite the contractor to do so as the cheapest possible solution for all concerned. The contractor's liability for such latent defects will be governed by the Limitation Act 1980 subject to whatever may be the conclusive effect of the final account and final statement (see section 10.6).

CHAPTER EIGHT
EXTENSION OF TIME

8.1 *Principles*

Under the general law and in the absence of any contractual provision empowering the award of an extension of time, the contractor is bound to complete the Works by the date agreed, unless he is prevented from so doing by some action or inaction of the employer. The position was neatly put by Lord Fraser of Tullybelton in *Percy Bilton Ltd* v. *Greater London Council* (1982):

> 'The general rule is that the main contractor is bound to complete the work by the date for completion stated in the contract. If he fails to do so, he will be liable for liquidated damages to the employer... That is subject to the exception that the employer is not entitled to liquidated damages if by his acts or omissions he has prevented the main contractor from completing his work by completion date... These general rules may be amended by the express terms of the contract.'

This general rule is amended in WCD 98 by clause 25 which expressly confers on the employer the power to extend the contract period for specific reasons. If there was no clause 25 and the employer prevented completion by the due date, or if the employer fails to make an extension of time as provided for under the clause, the contractor's obligation to complete by the contract date for completion is removed and his obligation is merely to complete within a reasonable time: *Wells* v. *Army & Navy Co-operative Society Ltd* (1902). Even if the contractor is subsequently delayed through his own fault, the employer cannot then deduct liquidated damages, because there is no fixed date from which the damages can be calculated: *Miller* v. *London County Council* (1934). If the employer's right to recover liquidated damages is to be kept alive, it is essential that the extension of time clause is operated properly and promptly, at least where the ground for the award is the 'fault' or responsibility of the employer. In that sense, the extension of time clause is for the benefit of the employer. Of course, it also benefits the contractor when it provides for an extension of time on grounds which are outside the employer's control and for which the contractor otherwise would have to take the risk. 'Exceptionally adverse weather conditions' is one such ground.

In the New Zealand case of *Fernbrook Trading Co Ltd* v. *Taggart* (1979), Mr Justice Roper took the view that, under the normal extension of time clause, a retrospective extension of time is only valid in two circumstances:

'(1) Where the cause of delay lies beyond the employer and particularly where its duration is uncertain ... although even here it would be a reasonable inference to draw from the normal extension clause that the extension should be given a reasonable time after the factors which will govern the engineer's discretion have been established.

(2) Where there are multiple causes of delay there may be no alternative but to leave the final decisions until just before the issue of the final certificate.'

In another New Zealand case – *New Zealand Structures and Investments Ltd* v. *McKenzie* (1979) – a different judge took the view that under the normal extension of time clause the certifier can grant an extension of time right up until the time he becomes *functus officio*, i.e. devoid of powers, which in most cases will be on the issue of the final certificate. The court said:

> 'In a major contract it is virtually impossible to gauge the effect of any one cause of delay while it is still proceeding, let alone assess the consequences of concurrent or overlapping causes. Finally, any need to have a prompt decision loses some force as a factor in interpreting such a clause, when one considers the normal review and arbitration procedures...'

This, in general, is a realistic approach.

It is commonly assumed by contractors that they must first secure an extension of time before they are entitled to claim direct loss and/or expense. Under traditional contracts, architects and quantity surveyors are often of the same mind. The sequence is often that the contractor first obtains an extension of time for various reasons. He then applies for direct loss and/or expense and the quantity surveyor is instructed to value the 'cost related' extensions at a figure per week extracted from the preliminaries to the bills of quantities or from actual costs. Many claims are settled quite amicably on this basis. However, it is not strictly correct. The contractor may have an entitlement to an extension of time, but not to loss and expense, or he may be entitled to loss and expense, but not to extension of time, or he may be entitled to extension of time and loss and expense. The obtaining of an extension of time is not a precondition to the recovery of loss and expense: *H. Fairweather & Co Ltd* v. *London Borough of Wandsworth* (1987). The misconceptions almost certainly arise from the fact that some of the grounds for extension of time are reflected almost word for word in clause 26 (loss and expense). The situation was explained in *Henry Boot Construction Ltd* v. *Central Lancashire New Town Development Corporation* (1980) by Judge Edgar Fay QC speaking about the very similar clauses in the 1963 JCT Standard Form:

> 'The broad scheme of the provisions is plain. There are cases where the loss should be shared, and there are cases where it should be wholly borne by the employer. There are also those cases which do not fall within either of these conditions and which are the fault of the contractor, where the loss of both parties is wholly borne by the contractor. But in the cases where the fault is not of the contractor the scheme clearly is that in certain cases the loss is to be shared: the loss lies where it falls. But in other cases the employer has to compensate the contractor in respect of the delay, and that category, where the employer has to compensate the contractor, should, one would think, clearly be composed of cases where there is fault upon the employer or fault for which the employer can be said to bear some responsibility.'

The effect of any delay is to be considered in relation to what is actually happening on site at the time the delaying factors operate, not in relation to what should have been happening on site by reference to any programme: *Walter Lawrence & Son Ltd* v. *Commercial Union Properties (UK) Ltd* (1984).

8.2 Contract procedure

Clause 25 closely follows the extension of time provision in JCT 98. There are some significant differences. The contractor must give notice in writing to the employer every time that it becomes reasonably apparent that progress is being or is likely to be delayed. That is made plain in clause 25.2.1. Common sense suggests that it is the contractor to whom it is to be reasonably apparent, because until he knows, he cannot act. The written notice must give the circumstances of the delay and, if the contractor considers that he is entitled to an extension of time, he must state the relevant event (clause 25.4) which covers the situation. The contractor must not simply notify delays for which he expects the contract period to be extended, but any delay to progress. For example, if an important piece of earth-moving machinery breaks down and takes three days to fix or replace, the contractor must report the fact to the employer. The idea is that the employer is kept fully informed of all factors which might result in the project completion being delayed, so that he can take action, for example, by issuing instructions to replace floor tiles with a more readily available product to make up for his own delays and to carefully monitor the results of the contractor's delays. If the contractor fails to give written notice of every delay, it is a breach of contract which the employer is entitled to take into account when considering an extension of time: *London Borough of Merton* v. *Stanley Hugh Leach Ltd* (1985) which also made clear that the giving of the notice is not a precondition to the award of an extension of time.

If the delay notified by the contractor is not identified as a relevant event, his duty ends there until another delay occurs. If, however, the delay is a relevant event, the contractor must give further information in the notice or as soon as possible afterwards. He must take each separate relevant event and estimate the effect on other items of work and the effect on the completion date. It may be that a delay, although it is a relevant event, has no effect on the completion date, possibly because it is not on the critical path of the project. If that is the case, the contractor must so state. Thus, the contractor may refer to three different relevant events and give the effects as two days, one week, three weeks respectively. That does not mean that the cumulative result will be equal to a simple aggregate, i.e. four weeks and two days. Some delays may have concurrent effects. It is quite difficult to isolate the effects in this way and many contractors use the computer to perform this chore. There are several excellent software packages which will allow the contractor to input his programme as a network and separately introduce delays and this approach has judicial approval: *John Barker Ltd* v. *London Portman Hotels Ltd* (1996). However difficult the task, the contractor must do his best. In cases where the situation is changing, the contractor must keep the information updated.

The employer's duties are set out in clause 25.3. He has 12 weeks in which to fix a new completion date. The 12 weeks is measured from receipt of the contractor's notice under clause 25.2.1 and reasonably sufficient particulars and estimates. If there are fewer than 12 weeks between the receipt of the information and the date for completion, he must make his decision before the date for completion. This may not be easy if the information is not available until only one week before completion date. In such circumstances, we believe his obligation is to fix a new date as soon as possible and 12 weeks from receipt of information would be unlikely to be accepted as reasonable in that situation. What constitutes reasonably sufficient particulars and estimates amounts to the information required under clause 25.2.2, i.e. in respect of each relevant event, the effects on other work together with an estimate of the effect on completion date. If the delay is continuing, it seems that the employer is not obliged to fix a new date until after the full details of the effects of the delay have been obtained. That date, therefore, should be capable of identification without too much difficulty.

There are three criteria which the employer must bring to the fixing of a new completion date:

(1) The date for completion must be fair and reasonable
(2) The event must be a relevant event (i.e. one of those listed under clause 25.4)
(3) The relevant event must be likely to result in the completion of the Works being delayed.

In *John Barker Ltd*, the judge's view was that extensions of time should be properly calculated and not made on the whim of the architect. In particular, the contractor's own delays should not be taken into account. We know of many architects who decide on extensions of time by taking the total time taken by the contractor to complete and then deducting what the architect feels to be the time for which the contractor is responsible. This kind of negative approach may have certain attractions, but it is definitely not what the contract requires.

It is possible that the Works are being delayed by circumstances which can be shown to fall under the head of one or more relevant events, but if it does not appear that the Works will be delayed beyond the date for completion set in the contract (or any previously extended date) the contractor is not entitled to any extension of time. The fixing of an extension of time by the employer is quite a difficult business. Since the employer or his agent is not under any duty to act fairly between the parties other than the contract duty to set a new date which is fair and reasonable, the employer's decision under this clause seems not in any sense to be binding unless the parties allow it to be binding.

The employer, however, has strict obligations so far as the notice fixing a new date is concerned. Apart from setting a new date, the employer must state which of the relevant events has been taken into account and the extent to which he has taken into account any instructions for omissions issued since the last extended date was fixed. It does not appear that the employer is under any duty to allocate the additional days extension alongside particular relevant events. It would be perfectly proper for him to fix a new date at, say, 9 September 1999 and to state that the following relevant events have been taken into account; 25.4.2, 25.4.5.1, 25.4.6 and 25.4.7. The contractor will usually be anxious to see a precise apportionment, because he will then use the 'cost related' relevant events to claim loss and/or expense.

The employer is entitled to take omission of work into account. Indeed, he may omit work specifically to avoid making a lengthy extension of time. In practice, once the construction process is under way, omission of work is unlikely to reduce the contract period to any significant extent unless the omission is itself significant. Clause 25.3.2 makes clear that after the employer has carried out the procedure once, he may take subsequent omissions into account to the extent of reducing a previously awarded extension of time. It seems that he may so act without the action being triggered by a notice from the contractor. For example, the contractor may submit a notice of delay following which the employer may fix a new date of 12 May instead of 15 April. Subsequently, the employer may issue a change instruction which has the effect of reducing the amount of work required of the contractor. The employer may then fix a new date for completion which is earlier than the 12 May so as to allow for the reduction in work. On no account can a time adjustment, required by the contractor and accepted under the provisions of clause 12.4.2.A7, be changed. Clause 25.3.5 stipulates that the contractor has no power to fix a date earlier than the date for completion noted in appendix 1.

The employer is obliged to review the extension of time after the date of practical completion. If practical completion occurs after the date for completion in the contract or as extended, the employer may carry out the review. A strict reading of clause 25.3.3 suggests that the employer may carry out the review only once, whether it is immediately after the date for completion or just before the expiry of the appropriate period after practical completion. This is a valuable power for the employer and it can prevent time becoming at large. In carrying out the review, the employer may consider omissions of work. He is not tied to any previous decision or notification by the contractor and he may act freely except that he must consider delays in relation to relevant events.

The employer has three options:

(1) He may fix a completion date which is later than the completion date already fixed
(2) He may fix a completion date which is earlier than the completion date having regard to his instructions requiring an omission issued after the last date on which time was extended
(3) He may confirm the completion date previously fixed.

He must act before the expiry of 12 weeks after the date of practical completion. It has been said that this period is merely directory, not mandatory – *Temloc Ltd* v. *Errill Properties Ltd* (1987) – and that the employer may take rather longer to make up his mind if he wishes. The views expressed in this case should be treated with care. The Court of Appeal was considering a situation where the employer was attempting to use to his own advantage his architect's failure to act. In our view, the court applied the principle that a party should not be allowed to profit through its own breach (*Alghussein Establishment* v. *Eton College* (1988)) and interpreted the provision *contra proferentem* against the employer. Where time periods are specifically set out in the contract, it is always prudent to adhere to them.

There is an important proviso in clause 25.3.4. The contractor must constantly use his best endeavours to prevent delay and he must do everything reasonably required to the satisfaction of the employer to proceed with the Works. The employer, when making an extension of time, is entitled to take into account the extent to which the contractor has complied with these requirements. The proviso does not empower the employer to require acceleration of the Works. If the parties agree that acceleration is possible and advisable, it must be the subject of a separate agreement. If they should embark on this course of action, however, the effect on other contract provisions must be considered. Among other things, clauses 16, 24, 25 and 26 may be affected and expert advice is required. Clause 25.3.4 appears to be nothing more than a duty to continue to work regularly and diligently. In the Australian case – *Victor Stanley Hawkins* v. *Pender Bros Pty Ltd* (1994) – it was defined as doing everything prudent and reasonable to achieve an objective. If the employer requires some action to be taken which does not involve the contractor in extra expenditure, but which may save some time, the contractor is obliged to comply.

8.3 *Relevant events*

The grounds which entitle the contractor to an extension of time are termed 'relevant events'. The employer may only fix a new date for completion if he is satisfied that the delay falls squarely within one of the relevant events under clause 25.4. The grounds comprise two categories: events which are attributable to the employer, and events which are attributable to neither employer nor contractor although they are not differentiated in the contract. Events falling into the first group are to be found under clauses 25.4.5, 25.4.6, 25.4.8, 25.4.12, 25.4.14, 25.4.16 and 25.4.17. Events falling into the second group are 25.4.1, 25.4.2, 25.4.3, 25.4.4, 25.4.7, 25.4.9, 25.4.10, 25.4.11, 25.4.13 and 25.4.15. The following descriptions are simplified; those falling into the first group are starred.

Clause 25.4.1 Force majeure

This is a term used in French law and it is broader than 'Act of God', referring to all circumstances independent of the will of man: *Lebeaupin* v. *R. Crispin & Co* (1920). It is seldom called in aid by a contractor suffering delay, because most of the circumstances which obviously fall under this clause are already covered under other relevant events. Such events as civil commotion, government decrees, fire or exceptional weather conditions spring to mind.

Clause 25.4.2 Exceptionally adverse weather conditions

This event is worded so as to embrace unusually dry as well as unusually wet or frosty conditions. A long hot summer can cause great problems on site, particularly with such matters as the curing of concrete. The key word is 'exceptionally'. The weather must be exceptionally adverse in the light of the kind of weather usually encountered at that time of year or in that place. The contractor is expected to allow for the normal deviation in weather patterns. In order to decide whether the weather fits that description, meteorological reports are helpful. It is suggested that reports for the previous ten years would be necessary to establish that the adversity was exceptional. In deciding whether a particular piece of exceptionally adverse weather is to be allowed as ground for an extension of time, the employer must consider its effect on the works at the stage they have actually reached, not the stage they should have reached in accordance with some programme. Even if the weather is affecting the Works because the contractor's own delay has prevented the building being sealed against the weather, the contractor is entitled to an extension of time if it appears likely the completion date will be exceeded as a result.

Clause 25.4.3 Loss or damage caused by specified perils

It is noteworthy that the contractor is not entitled to an extension of time for delay caused by all the insurance risks, but only the specified perils, which are noted as fire, lightning, explosion, storm, tempest, flood, bursting or overflowing of water tanks, apparatus or pipes, earthquake, aircraft and other aerial devices or articles dropped therefrom, riot and civil commotion, but excluding excepted risks, i.e. ionising radiations or contamination by radioactivity from any nuclear fuel or from nuclear waste from the combustion of nuclear fuel, radioactive toxic explosive or other hazardous properties of any explosive nuclear assembly or nuclear component thereof, pressure waves caused by aircraft or other aerial devices travelling at sonic or supersonic speeds.

The main difference between 'specified perils' and 'all risks' is the risk of impact, subsidence, theft or vandalism. Delay to the completion of the Works resulting from loss or damage caused by these risks must be dealt with by the contractor. This is a point which the contractor must take into account when required to obtain insurance cover under clause 22A (see section 11.3).

Clause 25.4.4 Civil commotion, strike, lock-out affecting trades employed on the Works, preparing, manufacturing or transporting materials for the Works or persons designing the Works

Strikes affecting three kinds of persons are included: persons working on site, persons working off-site getting things ready for site or delivering them, and persons designing the Works. It is uncommon for an independent consultant on the design team to have to deal with a strike; strikes in the other two categories are more likely. The relevant events must be read strictly. The strike provision is broad enough to encompass both official and unofficial strikes, but it will not include a work to rule. A strike affecting goods required for the Works is included, but not a strike affecting deliveries of raw materials to a factory for the manufacture of goods required for the Works.

*Clause 25.4.5 Compliance with employer's instruction under clauses 2.3.1, 12.2, 12.3, 23.2, 34 or regarding opening up and testing**

Clause 2.3.1 refers to divergences between the Employer's Requirements and the definition of the site boundary under clause 7 (see section 2.7). The employer must issue instructions and if compliance results in a likely delay to completion, an extension of time is indicated. These are the kind of divergences which usually arise at the setting out stage. In many cases, prompt instructions from the employer will result in virtually no delay or, at worst, a delay of just an hour or so. If the divergence is not discovered until after the Works are well advanced, the consequences in terms of delay may be very grave.

Clauses 12.2 and 12.3 refer to change instructions and instructions to expend provisional sums. Large numbers of small instructions may have effects which are quite out of proportion to their value compared to the total value of the Works. The practice of arriving at an 'appropriate' extension by simple proportioning of these values to times is quite mistaken. For example:

$$\frac{\text{Value of instructions: £50,000}}{\text{Contract sum: £500,000}} = \frac{\text{Extension of time: x}}{\text{Contract period: 10 months}}$$

Therefore the value of x = 1 month.

If there were 150 instructions to carry out all kinds of extra work, the delay might actually be much greater than one month. On the other hand, if there was only one instruction involving the pouring of many extra cubic metres of mass concrete, the delay might be very little. Indeed, it is possible that a large number of instructions, some to add, some to omit and others simply to change the work, may result in virtually no change in the contract sum, but the delaying effect may be severe. There are no firm rules and each instruction should be separately evaluated, not only in terms of money but also in terms of delay. The disorganising effect of a multitude of instructions should also not be overlooked.

An instruction for the expenditure of a provisional sum poses another kind of problem. A provisional sum can be included only in the Employer's Requirements (clause 12.3). Commonly, the information given about the subject of the provisional sum is sketchy. In many instances, the contractor cannot make any realistic provision in his programme and he is not obliged to guess any likely time. In such cases, the issue of an instruction may entitle the contractor to an extension of time which bears no relation to the difference between the provisional sum and the actual cost.

Clause 23.2 refers to the postponement of design or construction (see section 5.3). A postponement of three weeks in the execution of the Works midway through the project will certainly give rise to more than three weeks delay to the completion of the Works as the contractor has to carry out certain procedures before stopping (for example, make the Works safe) and to gear the workforce to start again. The effect of partial postponement will depend on the position of the postponed work in the network.

Clause 34 refers to the discovery of antiquities and the issuing of instructions to deal with them.

Opening up and testing of work is dealt with in clause 8.3. The contractor is entitled to an extension of time unless the work is not in accordance with the contract.

*Clause 25.4.6 Late instructions, decisions, etc. specifically applied for in writing neither too early nor too late**

Despite the wording, it has been held by a court considering a similar clause that the requirement for a specific application in writing is satisfied if the contractor produces a sufficiently detailed programme in which the dates on which information is required are clearly marked: *London Borough of Merton* v. *Stanley Hugh Leach Ltd* (1985). The contractor must update the programme information each time there is an extension of time. In traditional contracts, the equivalent clause normally provides the contractor with some of his grounds for extension of time. This type of procurement should reduce the opportunities for extensions under this head, because the contractor is in charge of the production of information necessary to construct the Works. This event will mainly apply if the employer is asked to give a decision on some matter which properly calls for his decision or consent. Clause 2.4.2 is expressly singled out by this event.

Clause 25.4.7 Delay in receiving necessary permission of a statutory body

This relevant event has already been considered in section 6.3. The contractor must have taken all practicable steps to reduce the delay. This event clearly refers to every kind of statutory permission or approval. Different buildings will attract differing regulations. Most buildings require planning permission and must satisfy the Building Regulations, but there are fire regulations, entertainment licences, water regulations and many more statutory controls which may apply. The contractor must be able to show that he has made any necessary applications at the appropriate time and that the reason for the delay is not something for which he is responsible, i.e. he has taken all practicable steps to reduce the delay. What 'practicable steps' are may appear to be self-evident. Useful guidance has been given in *Jordan* v. *Norfolk County Council* (1994) where the judge held that the term 'reasonably practicable' referred not just to physical practicability but also to whether a course of action was practicable in the financial sense.

*Clause 25.4.8 Work not included in the contract**

There are two subclauses to this relevant event. One deals with work and the other deals with materials carried out or supplied by the employer or by persons engaged by the employer. Clause 29 allows the employer to carry out work not included in the contract (see section 6.5). There is no clause in the contract which expressly provides for the employer to supply materials and it is noted that the subclause refers to materials and goods which the employer 'has agreed' to provide. In a situation where the employer was, say, a brick manufacturer, it could well be imagined that he would wish to supply bricks for the project. This could best be achieved by the inclusion of a suitable provision in the Employer's Requirements. It might be expected that the subclause would refer to materials and goods which the employer 'has elected' to provide. It should also be noted that the execution or failure to execute work and the supply or failure to supply materials are equally likely to entitle the contractor to an extension of time. An interesting dispute scenario would be set up if the employer agreed to provide materials which subsequently proved to be defective and caused a serious building defect.

Clause 25.4.9 Government exercise of statutory powers

This event applies only where the UK government has exercised powers after the base date. The exercise must directly affect the Works by restricting the availability of labour, materials or fuel which are essential to the carrying out of the Works.

Clause 25.4.10 Inability to secure labour or materials

There are two subclauses. One deals with labour, the other deals with materials. In order to qualify as a relevant event, the inability to obtain labour or materials must be for reasons which are beyond the contractor's control. In addition, he must have been unable to foresee such reasons at the base date. Although there will be times when a contractor will be unable to obtain certain materials no matter what measures he takes or what price he is prepared to pay, he will always be able to obtain labour. Sometimes he will have to pay a grossly inflated price or he may be obliged to bus them in to the site from some distance away, but he will always be able to secure labour. Since the times when the contractor will not be able to secure materials will be very rare, it might be asked, what is the purpose of this clause? In order to make sense of this particular event it is necessary to make some implication as to the availability of labour or materials at prices which could reasonably be assumed by the parties at the base date. To expect the employer to examine this event in that light is perhaps expecting too much and the criteria which differing employers and their agents would bring to the same facts are likely to vary widely. This is a peculiarly difficult event to consider in practice and in some forms it is optional. There is nothing to prevent the parties from striking the clause from this contract provided it is also remembered to strike out the appropriate clauses in the fluctuation provisions to preserve the 'freezing' of fluctuations at completion date (see section 10.9).

Clause 25.4.11 Work by statutory undertaker

This clause applies only when the work is being carried out in relation to the Works. It is quite possible that a statutory undertaker may delay one site, the access perhaps, while carrying out statutory obligations to lay pipe or cables to an adjacent site. That is not something for which the contractor would be entitled to an extension of time. This clause does not appear to apply to a statutory undertaker acting under its powers, if it is not acting as a duty. Neither does it apply to them when acting as directly employed contractors to the employer (in such a case, clause 25.4.8.1 would most likely apply).

*Clause 25.4.12 Failure to give access to site**

This clause is not the same as failure to give possession of the site.

The contractor is entitled to an extension of time under carefully restricted conditions:

- The access must have been laid down in the Employer's Requirements and the contractor must have given whatever notice was there specified; or the access must have been agreed between employer and contractor
- The access must be through or over land, buildings, way or passage adjoining or connected with the site
- The land, etc. must be in the possession and the control of the employer
- The employer must have failed to give ingress or egress at the appropriate time (i.e. at any time during the progress of the work).

The clause will bite only if all these conditions are satisfied. For example, if the employer has specified the access in his Requirements, but it is over land which is not in the employer's possession and the employer cannot secure such access, the contractor is not entitled to an extension of time. In that situation, the employer would be in breach of contract and an extension of time might well be irrelevant. If, however, access is specified from an adopted road and the access is blocked by roadworks, no extension of time is indicated and the employer is not in breach.

Clause 25.4.13 Change in statutory requirements

The change must have taken place after the base date and the change must have affected the Works as noted in clauses 6.3.1 or 6.3.2 (see section 6.3). Moreover, the contractor must have taken all practicable steps to avoid or reduce the delay.

*Clause 25.4.14 Deferment of possession**

If clause 23.1.2 is stated in appendix 1 to apply, the exercise by the employer of the power to defer possession for up to six weeks will inevitably attract an extension of time. The period of extension may, but will not necessarily, equal the period of deferment. It may be longer to allow for the remobilisation of labour and materials. This clause must be interpreted strictly. The employer has no power to extend time under this clause on failure to give possession if the deferment clause is not stated to apply or if the employer, having got the right, has not formally exercised it.

Clause 25.4.15 Use or threat of terrorism and the activity of authorities in dealing with it

Unfortunately, this relevant event needs no explanation.

*Clause 25.4.16 Compliance or non-compliance by the employer with clause 6A.1**

This relevant event will only take effect if the contractor does not take on the role of planning supervisor and principal contractor. It will be seldom that he does not assume the latter role, but the employer may often require the employer's agent to act as planning supervisor also. In such an instance, it should be noted that it is not simply the failure of the employer to ensure the carrying out of the planning supervisor's duties, but also the correctly ensuring which may cause delay and lead to extension of time. The planning supervisor's duties, properly carried out, will inevitably delay progress in some instances.

*Clause 25.4.17 Suspension of contractor's obligations**

If the contractor properly exercises his right under clause 30.3.8, he is entitled to an appropriate extension of time. What is appropriate is unlikely to be merely the length of the suspension. When the contractor receives payment in full, he will need time to get back to full production on site, depending on the stage the project has reached.

8.4 Liquidated damages

8.4.1 General principles

It is open to the parties to a contract to agree on a fixed sum of money which one will pay to the other in the case of a breach of contract. In the case of building contracts, it is usually stipulated that such a sum will be paid if the contractor fails to complete the Works by the date for completion in appendix 1. This contract is no exception and the terms of such payment are to be found in clause 24.

The arrangement saves the parties the uncertainty and expense of a legal action to determine the damages payable for the breach. In the absence of clause 24, the employer would be left to recover whatever amount of unliquidated damages he could prove he had suffered. The liquidated damages clause is often (incorrectly) referred to as a penalty clause. There is a significant difference between them. Liquidated damages must be a genuine pre-estimate of loss. That is to say that it must be a figure inserted into the contract by the employer to represent the best estimate he could make, at the date the contract was executed, of the likely loss he would suffer if the completion was delayed. A penalty, however, is a punishment whose value bears no relation to the damages expected to be incurred. A penalty is not enforceable. It makes no difference what terminology is used, it is the reality which is important. Certain guidelines have been set out for the recognition of a penalty, notably in *Dunlop Pneumatic Tyre Co Ltd* v. *New Garage & Motor Co Ltd* (1915) per Lord Dunedin:

> 'It will be held to be a penalty if the sum stipulated for is extravagant and unconscionable in amount in comparison with the greatest loss which could conceivably be proved to have followed from the breach... It will be held to be a penalty if the breach consists only in not paying a sum of money, and the sum stipulated is a sum greater than the sum which ought to have been paid... There is a presumption (but no more) that it is a penalty when a single lump sum is made payable by way of compensation, on the occurrence of one or more or all of several events, some of which may occasion serious and others but trifling damages... On the other hand ... it is no obstacle to the sum stipulated being a genuine pre-estimate of damage that the consequences of the breach are such as to make precise pre-estimation almost an impossibility. On the contrary, that is just the situation when it is probable that pre-estimated damage was the true bargain between the parties.'

In practice, many liquidated damages provisions are so badly expressed that they are either inconsistent with other clauses in the contract – *Bramall and Ogden Ltd* v. *Sheffield City Council* (1985) – or in operation they become penalties: *Stanor Electric Ltd* v. *R Mansell Ltd* (1988). If such a clause is held to be a penalty, all is not lost so far as the employer is concerned. He is left with his common law remedy of suing for such damages as he can prove. Although the point has not been conclusively settled, we are of the view that the employer would be unable to recover more than the amount set down as a penalty. Any other conclusion would be inequitable.

Liquidated damages are recoverable without proof of loss. It matters not that the employer has lost less than expected, that he has lost nothing at all, or even that he has made a profit as a result of the contractor's late completion. The employer is entitled to the liquidated damages in each of these instances: *BFI Group of Companies Ltd* v. *DCB Integration Systems Ltd* (1987). The amount of liquidated damages is to be inserted in appendix 1 as £... per... If the space is left blank, the employer would not be entitled to recover any liquidated damages, but it is probable that he could recover his actual loss on the same basis as if the sum indicated was a penalty except that there would be no ceiling on the possible recovery. If, however, 'nil' is inserted, that figure would signify the amount of damages recoverable per day or per week and the employer would be unable to sue for damages at common law, because the provision for liquidated damages is exhaustive of the employer's rights to damages: *Temloc Ltd* v. *Errill Properties Ltd* (1987).

8.4.2 Contract provisions

Clause 24.1 provides that if the contractor fails to complete the Works by the completion date or by any extended date, the employer must issue a written notice to the contractor. The notice is important. It is a pre-condition to the right of the employer to recover liquidated damages. It does not have the same weight as an architect's certificate under clause 24 of JCT 98 and the employer cannot rely on it so as to deduct liquidated damages from retention money so as to extinguish the fund: *J.F. Finnegan Ltd* v. *Ford Seller Morris Developments Ltd* (1991). If the employer makes a further extension of time after issuing the notice, clause 24.1 makes clear that he must cancel the original notice and issue a new one. This follows the judgment in *A. Bell & Son (Paddington) Ltd* v. *CBF Residential Care & Housing Association* (1989).

The contractor must pay or allow the liquidated damages and the employer may either deduct them from any monies due or to become due to the contractor or he may recover them as a debt. Not surprisingly, liquidated damages are invariably deducted. Following the provisions in the Housing Grants, Construction and Regeneration Act 1996 and in what seems to be an excess of caution, the recovery process has been made quite complicated. There are three pre-conditions which must be satisfied before recovery of liquidated damages can take place:

(1) The employer must have issued a notice of non-completion under clause 24.1; *and*

(2) The employer must issue a written notice to the contractor, informing him that he may require payment of, or may withhold or deduct the liquidated damages (clause 24.2.1); *and*
(3) The employer must issue a written notice or requirement requiring payment or notifying deduction (clauses 24.2.1.1 and 24.2.1.2).

The second notice may be served at any time after the non-completion notice, but not later than the date on which the final account and final statement become conclusive. Clause 24.2.3 provides that the employer need only serve one notice requiring payment. It remains effective, despite the issue of further non-completion certificates, unless the employer withdraws it. Since the decision to deduct liquidated damages rests with the employer, it is unlikely that he would ever, in practice, withdraw the notice. If he decided not to deduct damages, he would simply let the matter rest. However, after these two notices have been issued, the employer must do one of two things, depending on whether he intends to ask for payment of the liquidated damages or to deduct them from money due. This third notice must be issued not later than five days before the final date for payment under clause 30.6 (final account and final statement). This is obviously drafted to comply with legislation.

Some doubt has been thrown on the precise form to be taken by the employer's written requirement for payment. In the *Bell* case, considering a similar clause under the JCT 80 Standard Form, Judge John Newey QC stated:

> 'There can be no doubt that a certificate of failure to complete given under clause 24.1 and a written requirement of payment or allowance under the middle part of clause 24.2.1 were conditions precedent to the making of deductions on account of liquidated damages or recovery of them under the latter part of clause 24.2.1.'

This seems perfectly clear, but in *Jarvis Brent Ltd* v. *Rowlinson Construction Ltd* (1990), again considering JCT 80, it was held that the written requirement was satisfied by a letter, written by the quantity surveyor and forwarded to the contractor, which stated the amount which the employer was entitled to deduct; alternatively, it was stated that the cheques issued by the employer from which liquidated damages had been deducted constituted such written requirements. The judge went on to consider whether the written requirement was indeed a condition precedent and came to the conclusion that it was not. All that was necessary was that the contractor should be in no doubt that the employer intended to make the deduction. In *Holloway Holdings Ltd* v. *Archway Business Centre Ltd* (1991) a similar clause in IFC 84 was considered and it was again held:

> 'For (the employer) to be able to deduct liquidated damages there must both be a certificate from the Architect and a written request to (the contractor) from (the employer).'

The matter was finally clarified by a decision of the Court of Appeal in *J.J. Finnegan Ltd* v. *Community Housing Association Ltd* (1995) where the court held that the decision in *Bell* was correct and that the employer's written requirement was a condition precedent to the deduction of liquidated damages. Only two things must be specified in the requirement and they are:

- Whether the employer is claiming a payment or a deduction of the liquidated damages; *and*
- Whether the requirement relates to the whole or part of the total liquidated damages.

Clause 24 of WCD 98 is, of course, in very similar terms. It is our view that the current clause 24 places a duty on the employer to require the payment or allowance in writing before making a deduction of liquidated damages. Quite apart from the plain words of the contract (which also occur in JCT 98 and IFC 98), the new clause 24.2.3 emphasises that a requirement which has been stated in writing remains effective even if the employer issues further non-completion notices. We cannot understand why it is thought necessary for three notices to be given and we would not be surprised to learn that the intention was that two notices only were intended by the draughtsman. However, whatever the intention may have been, it is our view that, on the present wording, three notices are now essential before recovery of liquidated damages under the contract is permitted.

The amount which the employer may deduct is to be calculated by reference to the rate stated in appendix 1. The employer is free to reduce the rate, but not to increase it. That is expressly stated in clauses 24.2.1.1 and 24.2.1.2: 'or at such lesser rate . . .'. Clause 24.2.1 now makes clear that the employer need not wait until practical completion before deducting liquidated damages. He may start to deduct them as soon as he has issued the relevant notices. In practice, such deductions usually commence from the first payment thereafter.

In section 7.3 we discussed the complication introduced into practical completion by the requirement to comply with clauses 6A.5.1 and 6A.5.2. Although undoubtedly the parties are agreeing to take practical completion, for all the purposes of the contract (including the recovery of liquidated damages), as being the date specified in clause 16.1, it is possible that practical completion of the physical Works may take place days or even weeks before compliance of the relevant parts of clause 6A is finally achieved. Since compliance may be complete except for some relatively minor parts of the health and safety file, it is at least arguable that where physical practical completion has been achieved as a matter of fact and where there are minor parts of the health and safety file outstanding, the liquidated damages amount may be a penalty.

CHAPTER NINE
FINANCIAL CLAIMS

9.1 *Types of claim*

The dictionary definition of 'claim' is 'an assertion of a right'. All too often, claims in the building industry are looked on, and sometimes are, assertions of presumed rights which have no foundation in reality. In the context of building, a claim is usually a claim for money outside the contractual machinery for valuing the work. It may also be a claim for extension of time. There are three kinds of money claim commonly made by contractors:

(1) *Contractual claims* These are claims which are made under the express provisions of the contract. They are outside the normal contractual mechanism for valuation of work carried out and WCD 98 deals with them in clauses 26, 34.3 and supplementary provision S7. These clauses set out specific grounds on which the contractor can claim and they also specify the precise manner in which he must proceed with his claim. Many of the grounds would not allow the contractor to claim extra payment if they were not in the contract, because they are not breaches. In general, where the contract specifies certain requirements which must be satisfied in order to entitle the contractor to recover direct loss and/or expense, failure to satisfy those requirements will prevent recovery. These are the only kinds of claims which the employer's agent is empowered to deal with and the only kinds of claims which the employer may deal with under the contract.

(2) *Common law claims* These claims arise outside the express provisions of the contract. They may be claims in tort or for breach of the contract's express or implied terms. They may be claims that there is no concluded contract and that the contractor is, therefore, entitled to be paid on a quantum meruit basis. Nothing in WCD 98 prevents the contractor proceeding with any such claims through the dispute resolution procedure specified in the contract, provided the action is brought before

the final account and final statement become conclusive. Such claims are sometimes termed ex or extra contractual claims. The employer's agent has no power to deal with them. The employer may deal with them, of course, but not under the contractual claims procedures unless the contractor agrees.

(3) *Ex gratia claims* These are sometimes known as 'hardship' claims and they have no legal foundation. The contractor has no entitlement as a right. A contractor may advance this sort of claim as a last resort when he knows that he has suffered a large loss without there being any fault on the part of the employer. It is entirely up to the employer whether he meets this claim in full or at all and it is suggested that he will not normally do so unless there are special circumstances which make it advantageous for the employer to take this step.

Contractors who make claims are often unfairly dubbed 'claims conscious'. A contractor who makes justified claims is simply efficient and takes advantage of the procedures which are in the contract or in the common law precisely for the purpose of providing him with a remedy under the particular circumstances when it is right that he should have one. The contractors of which one should beware are those who make inflated or spurious claims as part of their normal approach to any project.

9.2 *Application for direct loss and/or expense*

WCD 98 sets out a procedure which the contractor must observe if he wishes to recover direct loss and/or expense under the terms of the contract. The contractor has no duty to carry out the procedure, but if he does not, he cannot recover loss and/or expense. If he wishes to set the procedure in motion, he must make application under clause 26.1. The application must be in writing and it must state that he has or he is likely to incur direct loss and/or expense for which he would not be reimbursed under any other provision of the contract. That is important. If the contractor is entitled to recover under some other provision, for example clause 12, he cannot recover under this clause.

The contractor must make the application as soon as it is apparent or should reasonably have become apparent to him that the regular progress of the Works has been or was likely to be affected. The contractor must act promptly. Although he may not be able to give details, he will very quickly know whether regular progress is likely to be affected by any particular circumstance. If he fails to act promptly, he will lose his entitlement to reimbursement of loss and/or expense under the contract.

The grounds upon which the contractor founds his application must be stated. They may be deferment of giving possession of the site (where clause 23.1.2 is stated in appendix 1 to apply) or regular progress being materially affected by one of the 'matters' listed in clause 26.2. Use of the phrase 'materially affected' makes clear that the contractor may only apply when the effect on progress is substantial. Trivial disturbances or delays must be absorbed. The degree of affection which can be categorised as 'material' may be disputed.

It should be noted that, in the 1998 edition of this contract, the criterion is that regular progress has been *and* is likely to be materially affected. On a strict reading of this clause it appears that the contractor cannot make application unless he makes it at a time when regular progress not only has been (i.e. in the past) materially affected but also will be affected in the future. In the previous edition, the italicised *and* above was expressed as 'or' so that a contractor could choose whether to apply after the material affection or before it. It should also be noted that a phrase has been inserted to make clear that the contractor is entitled to submit his own quantification of the direct loss and/or expense to which he believes he is entitled. We believe the insertion to be of little significance, because it does not place an obligation on the contractor to quantify and in our experience most contractors submit their own quantification as a matter of course.

What the contractor can recover as direct loss and/or expense is established as the same as could be recovered as damages at common law under the rules set out in *Hadley* v. *Baxendale* (1854). That is the loss which the parties could reasonably foresee would be the direct result of the breach of contract, considered at the date when the contract was entered into. There are no particular limits to the losses which the contractor can recover as direct loss and/or expense provided that they are within the reasonable contemplation of the parties and flow directly from the event relied on. What he can include will depend on the facts in each case.

Common heads of claim include the following:

- Plant and labour inefficiency
- Increases in cost
- Increases in head office overheads
- Acceleration

- Establishment costs during any period of delay
- Loss of profit (if the contractor can show that he could earn the profit elsewhere)
- Interest and financing charges: *F.G. Minter* v. *Welsh Health Technical Services Organisation* (1980).

Clause 26.1.2 provides that the contractor must support his application and the amount of loss and/or expense by giving the employer the information and details which the employer may reasonably require. It is to be noted that it is the employer's requirement, not the details, which are to be reasonable. The contractor is not obliged to provide supporting information unless it is specifically required by the employer. In *London Borough of Merton* v. *Stanley Hugh Leach Ltd* (1985), Mr Justice Vinelott famously said about the contractor applying for loss and/or expense:

> 'He must make his application within a reasonable time: it must not be made so late that, for instance, the architect can no longer form a competent opinion on the matters on which he is required to form an opinion or satisfy himself that the contractor has suffered the loss or expense claimed. But in considering whether the contractor has acted reasonably and with reasonable expedition it must be borne in mind that the architect is not a stranger to the work and may in some cases have a very detailed knowledge of the progress of the work and of the contractor's planning.'

The contract under consideration was JCT 63 with provision for an architect, and the loss and/or expense clause was somewhat different in wording, but if the employer has engaged an agent, it is reasonable to suppose that he will not be a 'stranger to the work' either. Whether the employer's request for any particular piece of information is reasonable should be considered in that light.

There is another difference between the subject of Mr Justice Vinelott's comments and this form. Under JCT 63, it is the responsibility of the architect to decide whether an application is valid and then to ascertain, or to instruct the quantity surveyor to ascertain, the loss and/or expense. In WCD 98, it is simply stated that the amount of the loss and/or expense incurred or being incurred by the contractor is to be added to the contract sum. Because it is the contractor who is to make application for interim payment under clause 30.3.1, he must also calculate the amount of loss and/or expense he is suffering. The employer's right to 'reasonably require' information is for the purpose of checking the contractor's application, not to ascertain it in the first instance. Read in this context, the insertion of a power, rather than a duty, for the contractor to quantify in clause 26.1 seems odd. The employer's agent is not charged with holding the balance between the parties, neither it seems does he owe the employer a duty to act fairly (see section 5.1). In this context, the ascertainment of loss and/or expense under WCD 98 has a changed emphasis. Whereas, under JCT 98 and most other standard forms in the traditional mould, it is the architect who determines the amount payable to the contractor as loss and/or expense and it is for the contractor to successfully dispute the amount if he can, under this form the burden of showing that the contractor's ascertainment is wrong is laid on the employer. This gives the contractor a distinct advantage.

The contractor, however, must be able to show how he has incurred the loss and/or expense in some detail. It is not sufficient for him to list contractual grounds in a general way ('We have been delayed and disrupted due to extra work, lack of approvals and interference by other contractors'). The disruptive matter must be specified and the effect noted in each case: *Wharf Properties Ltd* v. *Eric Cumine Associates* (1991). Where the consequences of the various matters interact in a complex way, it may be difficult or impossible to separate the evaluation. In that case, the contractor is entitled to put forward a composite calculation: *J. Crosby & Sons Ltd* v. *Portland Urban District Council* (1967). There is considerable misunderstanding of this point in the industry. In simple terms, the causes and effects must be individually identified, but in certain circumstances, the calculation of resultant loss and/or expense may be carried out on a global basis: see *ICI Chemical Industries plc* v. *Bovis Construction Ltd* (1992).

In that case, ICI brought proceedings against Bovis, the management contractor, as well as against the architects and consulting engineers. ICI's statement of claim contained several pages of allegations against each of the defendants, but these were not otherwise particularised and no attempt was made to link any alleged breach to a particular loss. In paragraph 21 a global claim was made against the defendants for a sum of £19 million and professional fees.

The pleadings were amended but still not it seemed – at least to the defendants – adequately particularised. Judge Fox-Andrews ordered ICI to serve a Scott Schedule setting out the alleged complaint, against whom it was made, which clause of the contract had been breached and the alleged factual consequences of each breach.

They complied with his order but still not to the satisfaction of the defendants; many of the items were pleaded on a 'global' basis.

This practice first received the approval of the Courts in *Crosby* v. *Portland* and was blessed by Mr Justice Vinelott in *Merton* v. *Leach*. The global approach is only acceptable, it has been said, where a claim depends on 'an extremely complex interaction in the consequences of various denials, suspensions and variations, [where] it may be difficult or even impossible to make an accurate apportionment of the total extra cost between several causative elements'.

Under the 'global approach' there must (self-evidently) be no duplication and, as was noted in *Leach*, a global award:

> 'can only be made in a case where the loss or expense attributable to each head of claim cannot in reality be separated and ... can only be made where apart from that practical impossibility the conditions which have to be satisfied before an award can be made have been satisfied in relation to each head of claim'.

It is not an excuse for sloppy pleading or for failure to prove one's case as was emphasised by the Privy Council in *Wharf* v. *Cumine* although their Lordships expressed no reservations about the correctness of the 'global approach'. What the Privy Council actually said was:

> 'What those cases actually establish is no more than this, that in cases where the full extent of extra costs incurred through delay depends upon a complex interaction between the consequence of various events, so that it may be difficult to make an accurate apportionment of the total extra costs, it may be proper for the arbitrator to make individual financial awards in respect of claims which can conveniently be dealt with in isolation and a supplementary award in respect of the financial consequences of the remainder as a composite whole. *This, however, has no bearing upon the obligation of a plaintiff to plead his case with such particularity as is sufficient to alert the opposing party of the case which is going to be made against him at the trial.*' [our italics]

Having surveyed these cases, Judge Fox-Andrews considered the Scott Schedule in detail. He found some of the items 'objectionable' and another 'hopelessly inadequate'.

> 'In respect of the many hundreds of items itemised on 101 pages, the financial consequences of which are always stated to be the

> same, namely "The total cost of abortive work amounts to £840,211 as particularised in Appendix 7..." ICI's case was that the various events set out all contributed to the sums claimed, with no actual apportionment being possible ... I find that it is palpable nonsense that £840,211 would be the cost of repositioning a bell. It is important to appreciate that whilst a pleading may take a particular form where a number of interactive events give rise to delay and disruption, the same does not appear to me to apply to many of the items [listed by ICI]. *The financial consequences of each breach, where possible, must be pleaded and the necessary nexus shown*'. [our italics]

In the event, the learned judge decided that since a great deal of work had been done by ICI and their advisors, they should not be debarred from pursuing their claim. However, he ordered that a fresh Scott Schedule giving the necessary particulars and showing the causal nexus should be served. A totally new and revised document was required.

If it is possible to evaluate the delaying or disruptive effects of individual causes this must be done, leaving only the balance of the delays for which the employer is alleged to be responsible to be swept up by the 'global approach'. It is clear that the contractor may put his claim in whatever form he wishes, but claims on a 'global' basis may suffer severe evidential problems: *GMTC Tools & Equipment* v. *Yuasa Warwick Machinery* (1994). The judgment in *How Engineering Services* v. *Lindner Ceilings Partitions plc* (1995) is very instructive on this point. There, in a careful judgment which is of general application, the court said that the claim must be intelligible and it must identify the loss, the reason for it and why the other party has an enforceable obligation to compensate for the loss. The claim should tie breaches to contract terms which should be identified. Cause and effect should be linked. Although there is no obligation on the contractor to break down the loss to identify the sum claimed for each specific breach, failure to do so will create an 'all or nothing' claim which will completely fail if some of the events cited are not substantiated. The court concluded by stating that a global claim must identify two things. The first was a means by which the loss is to be calculated if some of the events are not established. There should be some kind of realistic formula to achieve the scaling down of the claim. The second thing was a means of scaling down the claim to take account of the various other factors such as defects, inefficiencies or events which are at the contractor's risk. The calculation of loss should be carried out accordingly.

In clause 26.1, the words 'which has been or *is being* incurred' (our italics) indicate that the contractor need not wait until he has finished incurring loss and/or expense before it 'shall be added to the Contract Sum'. The matter is put beyond doubt by clause 26.3 which states that amounts ascertained 'from time to time' must be added to the contract sum. Clause 3 states that as soon as any amount is ascertained 'in whole or in part' it must be included in the next interim payment. The scheme of clause 26 is straightforward. Briefly, it is for the contractor to apply as soon as he realises that he is suffering loss and expense. He must include the appropriate sums, as they can be ascertained, in applications for payment and if the employer requires supporting information, it must be provided. The contractor is entitled to have the sums included in the next payment so that he is not unreasonably kept out of his money.

Clause 26.4 states plainly that the provisions of clause 26 are without prejudice to other rights and remedies which the contractor may possess. This means that the contractor may choose to exercise his common law rights instead of relying on clause 26, or he may apply under clause 26 in order to recover whatever he can and 'top up' this amount later with a common law claim. It also means that the contractor is not tied to the times and procedures set out in clause 26 except in one important respect. Most claims at common law will be based on the employer's alleged breaches of express or implied terms of the contract. Several of the grounds set out in clause 26 are not breaches of contract. For example, it cannot be a breach of contract for the employer to issue an instruction requiring a change or opening up of the Works, because the contract gives the employer power to issue just such an instruction. Therefore, if the contractor does not satisfy the provisions of clause 26 in respect of grounds like these which are not breaches of contract, he will be unable to bring a successful action for damages on these grounds at common law.

A contractor who intends to bring a common law claim should also note the provisions of clause 30.8.1 (see section 10.6). When the final statement has become conclusive under the terms of the contract, it is conclusive evidence, among other things, that where the contractor has received payment under the provisions of clause 26, in broad terms it finally settles any claims he has or may have in the future. There are some significant points to note about clause 30.8.1.3 which contains this provision:

- The finality is expressed as referring to claims for breach of contract, duty of care (tort), statutory duty or otherwise. It is suggested that 'otherwise' would be interpreted *ejusdem generis* in this instance to refer to claims of the same class as those expressly mentioned.

- The finality refers only to claims arising out of the occurrence of any matters in clause 26.2. Therefore, it is open to a contractor to bring a subsequent common law claim concerning some breach, etc. which is not included in clause 26.2. The opportunities for such claims are limitless. For example, breach of most of the employer's obligations under the contract are not included in clause 26.
- The inclusion of the words 'if any' after reference to the reimbursement of direct loss and/or expense restricts the operation of this clause to those circumstances where there has been some payment under clause 26. It appears that if there has been no payment under clause 26, the contractor is free to pursue his common law claims after the date at which the final statement becomes conclusive. A contractor wishing to bring a common law claim must be careful, but clause 30.8.1.3 is probably not nearly so final as is generally thought.

9.3 *Supplementary provision procedures (S7)*

These procedures greatly resemble the procedures in the Association of Consultant Architects Form of Building Agreement (ACA 2). They are very straightforward and sensible and promise advantages to both parties if properly operated. The first thing to note is that clause 26 is modified, but not superseded, by this provision. The second thing is that loss and/or expense which is dealt with under clause S6 (valuation of change instructions) is excluded from treatment under this clause. That is simply to avoid any possibility of the contractor obtaining double recovery.

The procedure is triggered as soon as the contractor is entitled to have some direct loss and/or expense added to the contract sum. He must include an estimate of the amount in his next application for payment. The amount must refer to the period immediately before the application. Therefore, if the contractor is receiving payments at monthly intervals, he must include in his estimate the whole of the amount he requires to represent his direct loss and/or expense during that month preceding payment. This places an obligation on the contractor to act swiftly when he becomes aware that he is incurring losses. Besides making the normal application in accordance with clause 26.1, he must calculate the loss and/or expense and insert it into the next payment application (clause S7.2). It is termed an 'estimate' rather than an ascertainment, because in many instances the contractor will not be able to calculate a precise figure.

In some cases, the contractor will incur the loss and/or expense over a long period. Clause S7.3 stipulates that the contractor must continue to submit estimates for as long as necessary, each estimate referable to the preceding period. Therefore, in the example noted above, the contractor would submit estimates every month until the loss ended and each estimate would refer to the preceding month (clause S7.3).

The contractor's estimates are dealt with in accordance with clause S7.4. The employer has 21 days from receipt of the contractor's estimate in which to give a written notice to the contractor. The employer may request information reasonably required to support the contractor's estimate, but he must request and receive such information within the 21 days. He may not delay giving notice on the ground that he has not received information. Of course, the content of the employer's notice will doubtless depend very much on the information received. The employer may not simply reject the contractor's estimate, but he may state one of the following options in his notice:

- He may accept the estimate.
- He may state that he wishes to negotiate the amount. No time limit is given for the negotiations and it is suggested that it is in the interests of the parties to insert a short time limit. Seven days is not unreasonable. If the employer chooses this option, he must also choose one of two further options to be put into effect if agreement cannot be reached. He must either refer the issue for a decision under article 6A (arbitration) or 6B (legal proceedings).
- He may simply state that the provisions of clause 26 apply.

The contract is silent as to what follows if the issue is referred under articles 6A or 6B. However, if agreement is reached on the amount of the addition to the contract sum, the sum must be added and no further sums may be added for that particular matter in clause 26.2 and in that period. There is an element of rough justice here in that both contractor and employer are held by tight time restraints and the contractor's finally agreed estimates may be rather wide of the mark. The idea is clearly that he receives payment as soon as practicable after the event and in return the employer has the certainty that there will be no more claims for the effects of any clause

26.2 matter during that period. How does it work in practice? Assuming that the contractor submits his applications for payment every four weeks, the contractor will submit his estimate with his general application for payment under clause 30.3.1. The employer has only 14 days in which to make payment following an application (clause 30.3.6) so the contractor will not receive payment then, because the employer has 21 days to issue his notice. If he issues the notice on the last of the 21 days, if he opts to negotiate and if there is a time limit of seven days imposed, agreement should be reached (or not reached) by the date of the next application and payment of the estimate can be made 14 days thereafter. If the employer opts to revert to clause 26, the contractor is not disadvantaged, because he is simply in the same position as he would be if clause S7 was not in the contract so far as that particular claim is concerned. The contractor is not then precluded from taking a reasonable time to calculate his precise loss and/or expense and to add the extra amount to his next application. As can be seen from clause 26, the employer is not in the position of ascertaining the amount due to the contractor, but if he disputes it, he must prove on the balance of probabilities that the contractor's ascertainment is wrong.

There is a sting in the tail (clause S7.6). If the contractor fails to submit his estimates as required under clauses S7.2 and S7.3, clause S7 ceases to apply. The contractor's direct loss and/or expense is to be dealt with under clause 26, but the amounts are not payable until the final account and final statement are agreed. Under these circumstances, the contractor is not entitled to any interest or financing charges incurred before the issue of the final account or final statement. It is, therefore, important for the contractor to rigidly observe the rules laid down in this clause if he wishes to secure maximum advantage. Theoretically, the contractor should have few claims under this form of contract, because he has complete control over the design and construction. Unfortunately, experience shows that employers cannot resist changes during the progress of the Works as well as the appointment of direct contractors to undertake special work. The result can only be justified loss and/or expense claims from the contractor.

9.4 *Grounds for direct loss and/or expense*

The grounds on which the contractor may make application for direct loss and/or expense are contained in clause 26. One is included in clause 26.1 and the remainder are referred to as 'matters' and they are contained in clause 26.2. In many instances, they echo relevant events in clause 25.4. This has doubtless led to the erroneous view that it is necessary for the contractor to obtain an extension of time under the appropriate clause before he is entitled to apply for loss and/or expense resulting from the same occurrence. That this approach is wrong is clear from the judgment in *H. Fairweather & Co Ltd* v. *London Borough of Wandsworth* (1987) when this question was considered in relation to JCT 63 and Judge Fox-Andrews QC said:

> 'But I do not consider that the obtaining of an extension of time under (the extension of time clause) is a condition precedent to recovering loss and expense under (the loss and expense clause)'.

The grounds are as follows.

Clause 26.1 Deferment of possession

Clause 23.1.2 must be stated in appendix 1 to apply or the employer has no power to defer possession. If it is not stated to apply, but the contractor has not received possession on the due date, the employer is in breach of contract and the contractor is entitled to damages at common law and probably to treat the contract as repudiated if the delay continues for a substantial period. Recovery cannot be made under the contractual machinery in those circumstances. Deferment of possession will obviously cause delay. To what extent it also causes the contractor to incur direct loss and/or expense it is for the contractor to demonstrate.

Clause 26.2.1 Opening up and testing

The employer's power to order opening up and/or testing is contained in clause 8.3 (see section 5.3.2). In order to qualify as a ground for recovery of direct loss and/or expense, the work, materials or goods must be shown to be in accordance with the contract. In such cases it is almost inevitable that the contractor will have a claim. In typical circumstances, there will be an instruction under clause 8.3 stipulating that the employer's agent must be present at the opening up. When the contractor opens up there will be a pause to allow the parties to decide whether the work conforms to the contract. It may take some time before this stage is completed especially if tests have to be taken to clarify the point. If it is decided that the work does conform, the disturbed work must be made good and only then can any work in that area proceed. It is clear that the decision to order opening up under clause 8.3 must be taken with a degree of circumspection related to the importance of the possible defect, its position in the Works and the effect of the instruction on progress.

Clause 26.2.2 Delay in receiving development control permission

The permission must be necessary for the Works to be carried out or to proceed and the contractor must have taken all practicable steps to avoid or reduce the delay. This probably means no more than that the contractor must make whatever application is necessary for permission as soon as he can. Development control requirements have been discussed in section 6.3. Although broader in meaning than merely planning requirements, they are certainly not as wide in meaning as statutory requirements of which they form part. Delays attributable to statutory requirements in general may perhaps be caught by clause 26.2.6 if they relate to a change in the requirements after base date as described in clause 6.3.1. In most instances, claims will be made under this head when the obtaining of planning permission and the like has been left to the contractor. The contractor cannot seek permission until after he has secured the contract and at that stage it is difficult to avoid delay. It is no doubt possible to delete the contractor's entitlement as a result of delay in obtaining permission, but it should be noted that several clauses interrelate on this topic and there is great danger of failing to completely deal with the matter: *Update Construction Pty Ltd* v. *Rozelle Child Care Centre Ltd* (1992).

Clause 26.2.3 Work not included in the contract

There are two subclauses to this ground. They are identical to the subclauses to clause 25.4.8 and the comments on that clause in section 8.3 apply equally to this clause.

Clause 26.2.4 Postponement instructions

Under clause 23.2, the employer may issue instructions to postpone any design or construction work (see section 5.3). It has been held that under the provisions of JCT 63 the architect may issue instructions which are effectively postponement instructions although he did not issue them expressly pursuant to the postponement clause: *Holland Hannen & Cubitts (Northern) Ltd* v. *Welsh Health Technical Services Organisation and Another* (1981). This possibility must not be discounted under this form of contract (substituting employer for architect), but unlike JCT 63, the contractor is unable to recover under this ground for such instructions. Only employer's instructions 'issued under clause 23.2' can be considered. For the equivalent of the *Holland* situation, the contractor might argue that all postponement instructions must be pursuant to clause 23.2 or they are not empowered under the contract and the contractor cannot comply.

Clause 26.2.5 Failure to give access to site

The wording of this clause is identical to the wording in clause 25.4.12 and the comments in section 8.3 are relevant. In the event that no access is shown in the Employer's Requirements, it would be for the employer and the contractor to agree access appropriate to the Works. It is likely that a term would be implied into every contract that the employer must give sufficient access to the contractor to enable him to carry out and complete the Works.

Clause 26.2.6 Instructions under clause 12.2 effecting a change or under clause 12.3 regarding the expenditure of a provisional sum

Any instruction under clause 12.2 may have financial repercussions beyond the ones provided for in the valuation rules. When that is the case, the contractor can claim under this head. This is also the head under which the contractor may be able to make application following a clause 6.3.1 situation where a change in statutory requirements after base date is to be treated as an instruction of the employer requiring a change. Where supplementary provision S6 applies, all the effects of the instruction are to be included in the contractor's estimate. The same is true where the contractor submits a priced statement under clause 12.4 alternative A (see Chapter 10).

Potentially the most significant are the employer's instructions regarding expenditure of a provisional sum. Such sums must be included in the Employer's Requirements, but they are often given with little explanation. The contractor has an obligation to use reasonable care in programming his work so as to achieve completion by the due date and it may be thought that he must take the work in provisional sums into account so far as he is able. Realistically, he will seldom be able to judge the work in a provisional sum with any accuracy and he is entitled to make little or no allowance in those circumstances. Indeed, it appears to us that under this form of contract the contractor is under no duty to take the work in provisional sums into account at all. The result is probably that an instruction to expend a provisional sum will have effect as a simple addition of work and/or materials.

Clause 26.2.7 Late instructions, decisions, etc. specifically applied for in writing neither too early nor too late

This clause is identical to clause 25.4.6. The omission of the word 'and' after 'clause 2.4.2' on the third line has been corrected in this edition. The comments in section 8.3 are relevant.

Clause 26.2.8 Compliance or non-compliance by the employer with clause 6A.1

This clause is identical to clause 25.4.16 and the comments on that clause apply here also.

Clause 26.2.9 Suspension of contractor's obligations

This matter refers to the same ground as the equivalent clause 25.4.17, but the wording is not identical. It should be noted that in order to recover loss and/or expense, the suspension must not be frivolous or vexatious. That immediately begs the question whether an extension of time may be made if the suspension is frivolous or vexatious. Clause 25.4.17 is merely putting into effect a legislative provision. There is nothing in the legislation to suggest that the contractor may suspend frivolously or vexatiously and the same can be said of clause 30.3.8. The introduction of this quite unnecessary proviso can only serve on occasion to provoke dispute.

9.6 *Antiquities*

Clause 34 deals with the situation if the contractor finds any fossils or other objects of interest or value and provides that the employer may give instructions regarding the examination, excavation or removal of the object. In any event, the contractor must use his best endeavours not to disturb the object and to preserve it in its exact position even if this involves cessation of work in whole or in part. Clause 34.3.1 provides that if the contractor suffers loss and/or expense as a result of complying with the employer's instructions, it is to be added to the contract sum. The only stipulation is that the contractor would not be reimbursed under any other clause. There is no requirement that the contractor must make application or supply information, and recovery under this clause is not linked to clause 26. Indeed, it appears deliberately not to have been so linked, because it would have been a relatively simple task to include antiquities as one of the 'matters' under clause 26.2. The contractor merely has to calculate his loss and/or expense and to submit it with his application for payment under clause 30.3.1. If supplementary provision S7 applies, the contractor must include these amounts in his claims under S7. Clause 34.3.2 provides that the employer must state in writing what extension of time has been made under clause 25.4.5.1 with reference to clause 34. He must do this to the extent necessary for ascertainment of loss and/or expense. In our view, there is no excuse for the inclusion of this provision which appears to give contractual sanction to the erroneous belief that there is a connection between extensions of time and loss and/or expense. There should be no such connection. The words '... to the extent that it is necessary for the ascertainment ...' mean that the employer will never have to take action under this clause: *Methodist Homes Housing Association Ltd* v. *Messrs Scott & McIntosh* (1997).

CHAPTER TEN
PAYMENT

10.1 Contract sum

WCD 98 is a lump sum contract. That is to say, the contractor carries out completion of the design and the whole of the construction for a stated amount of money which is to be paid by the employer. The fact that there are clauses which allow the amount to be varied is irrelevant. The important point is that the original sum is specified as being for a given amount of work. This is to be contrasted with a contract which expressly provides for remeasurement.

Clause 1.3 defines the 'contract sum' as 'the sum named in article 2 but subject to clause 14.2'. Article 2 contains a blank space in which the parties are to insert the figure on which they agree and clause 14.2 simply notes that the term 'contract sum' is exclusive of VAT. Clause 13 importantly stipulates that the contract sum must not be adjusted or altered in any way except in accordance with the express provisions of the contract. So, for example, a contractor who is having financial difficulties cannot call in aid an implied term allowing any such adjustment. The particular provisions of the contract which do allow such adjustment are noted in Fig. 10.1. Clause 3 provides that where it is stipulated that any amount is to be added to or deducted from the contract sum, or if an adjustment is to be made to the sum, as soon as ascertainment of the whole or part of the amount has taken place, it is to be included in the next payment. This clause is virtually identical to the provision in JCT 98, but it assumes particular importance in the absence of valuation by a quantity surveyor.

In order to understand the way in which the contract works it is important to remember that the value of the contract sum never changes. It may be adjusted, but it then becomes the contract sum which has been adjusted and in no sense is it a new contract sum. Article 2 is very clear about this. A sum is to be inserted and that sum is 'hereinafter referred to as "the Contract Sum"'. This is stated to be the sum which the employer will pay to the contractor 'or such other sum as shall become payable hereunder . . .'. Once the contract

Fig. 10.1
Adjustment of the contract sum.

Clause	*Adjustment*
2.3.1	Divergences between Employer's Requirements and definition of the site boundary.
2.4.1	Discrepancy within the Employer's Requirements.
3	Contract sum adjustments.
6.2	Fees legally demandable by Act of Parliament, etc. *and* stated by way of a provisional sum.
6.3	Change in statutory requirements or decision of the development control authority after the base date.
8.3	Opening up and testing.
9.2	Royalties and patent rights.
12	Changes.
13	Adjustment of contract sum.
16.2 & 16.3	Defects, shrinkages and other faults.
22B.2	Contractor insuring if employer defaults.
22C.3	Contractor insuring if employer defaults.
26.3	Loss and/or expense.
30.5.2	Final adjustment of the contract sum.
34.3.1	Loss and/or expense due to antiquities.
35–38	Fluctuations.
S6.4	Agreement of estimates.
S7.5	Direct loss and/or expense.

sum is adjusted it becomes 'such other sum'. We emphasise this apparently obvious point because we have recently become aware that certain construction professionals are confused by the concept.

Provisions for dealing with inconsistencies within the Employer's Requirements and the Contractor's Proposals are given in clause 2.4, but there are no provisions for dealing with errors in the contract sum or within the Contract Sum Analysis. Where the employer has, unusually, included bills of quantities in the Employer's Requirements and supplementary provision S5 applies, any errors in description or quantity must be corrected by the employer and treated as if they were changes in the Employer's Requirements. If the contractor makes an error in tendering by overlooking items or by making a mistake in adding up totals, he must bear the cost of the error himself. This type of procurement is very unforgiving in that respect and the employer and his professional advisors are unlikely to discover such errors before acceptance of the tender or indeed at any time, because of the nature of the procurement system. If, however, the employer discovers an error which is fundamental to the terms of the contractor's offer, such as the accidental omission of the fluctuations clause, the employer may not accept the offer so as to create a binding contract: *McMaster University* v. *Wilchar Construction Ltd* (1971).

10.2 *Interim payments*

A significant difference between this form of contract and others in the JCT series is that there is no provision for the quantity surveyor to value work carried out by the contractor and there is no provision for certification by an architect of money due to the contractor. The scheme of payments is basically quite simple. Clause 30.1 states that interim payments must be made by the employer to the contractor in accordance with clause 30.1 to 30.4 and either alternative A or B as stated in appendix 2. The amount in the interim payment must be the gross valuation as applicable to A or B less only the retention, any amount of advance payment due for reimbursement (see section 10.7) and any previous payments. Alternative A refers to stage payments and alternative B refers to periodic payments. They are fundamentally different as explained below.

10.2.1 Stage payments – alternative A

Where the employer wishes payment to be made in accordance with a series of stages, he must state this in the Employer's Requirements. The printed form contains a table in appendix 2 which the employer is to complete. It is a very simple table consisting of a column in which a brief description of each stage is to be inserted and another column in which the appropriate cumulative value of the stages is to be inserted. The cumulative value of the final stage is to be equal to the contract sum. Although the descriptions of the stages are to be brief, they must be clear enough so as to positively identify the stage in question. Thus, if isolated blocks are to be the stages, it will probably be enough to note: 'block 1, block 2, block 3', etc. and a final stage for the external works. If, however, the stages are physically connected, it is vital that the description precludes any dispute over the precise point at which it can be said that any particular stage has been completed. Although it may seem self-evident, periods of time are not to be inserted as stages. Thus 'four weeks, eight weeks, twelve weeks', etc. will certainly cause the contractor to apply for the stipulated amount after the appropriate lapse of time irrespective of whether he has actually completed an adequate amount of work. The employer could have difficulty in avoiding payment in such circumstances. We have encountered instances where the stages have been completed in this way and disputes have resulted.

In the previous edition of this form, it was open to the employer to include in the cumulative value the value of materials or goods stored off-site. In such an instance, note was to be made in the stages column and the value of the materials and goods had to be included in the cumulative values. We criticised this arrangement, because it was difficult to envisage how a stage could be complete if some of the materials were stored off-site. We assumed it was intended that the value would include the value of a completed stage plus the value of off-site materials stored for a future stage. Thus, a particular payment would include, in effect, a part payment of a future stage. This was workable, just, because the cumulative nature of the stages ensured that the payment for off-site materials was absorbed into future stage payments. We identified a particular problem which could occur. In order for the employer to include the value of off-site materials in a cumulative value, he must have made some estimate of the amount of materials which would be so stored at the time that the value column was completed. When the project was in progress, the materials might not actually be delivered as scheduled. This would have left the employer in the position of being liable to pay the appropriate cumulative value which would have included the value of materials not then stored off-site. To overcome this difficulty, the employer would have been obliged to operate the deduction provisions of clause 30.3.4 (see section 10.4). That was an inelegant solution, but the only one which we could devise to cope with the situation. By removing materials off-site from the cumulative value and inserting the amount into the list of amounts to be aggregated, this problem has been overcome.

Off-site materials and goods are now dealt with in clause 15.2. All such materials must be listed by the employer and the list must be supplied to the contractor and attached to the Employer's Requirements. The list is divided into two categories:

(1) Uniquely identified items such as lift equipment, purpose made furniture or special boilers; *and*
(2) Items which are not uniquely identified such as bricks, blocks, timber, tiles, etc.

The employer need not pay for any materials stored off-site unless they are listed. Fortunately, the employer can point to the lengthy proviso which lays down specific conditions which must be fulfilled before payment for off-site materials can be included. They are:

- *The contractor must provide reasonable proof that he owns the uniquely identified items and the contractor must have provided a bond in favour of the employer if so required in appendix 1.* The contractor may be hard pressed to provide reasonable proof in some instances. It appears that sight of the appropriate supply or subcontract terms together with proof of payment by the contractor would be acceptable for the ownership part of this condition. The employer may decide whether a bond is required, but if so, it must be from a surety approved by the employer and the terms should be those agreed between JCT and the British Bankers Association and attached to appendix 1 unless the employer requires the bond to be on other terms. The contractor must have been given a copy of such terms (sixth recital).
- *The contractor must provide reasonable proof that he owns items which are not uniquely identified and the contractor must have provided a bond in favour of the employer as noted in appendix 1.* The contractor may again be hard pressed to provide reasonable proof. Sight of the appropriate supply or subcontract terms together with proof of payment by the contractor would probably be acceptable for the ownership part of this condition. The bond is required and it must be from a surety approved by the employer and the terms should be those agreed between JCT and the British Bankers Association and attached to appendix 1 unless the employer requires the bond to be on other terms. The contractor must have been given copies of such terms (sixth recital).
- *The materials and goods must be in accordance with the contract.* This really is a superfluous condition. Materials and goods not in accordance with the contract can be rejected under clause 8.4.

- *The materials and goods must be, and remain, set apart at the place of storage or they must be clearly and visibly marked, individually or in sets by letters, figures or by reference to a pre-determined code so as to identify the employer and the destination must be marked as being the Works.* The main purpose of this condition is to make clear that these particular goods have been set aside as the property of the employer. In the case of a liquidation of the main contractor or another firm on whose premises they are stored, the employer would be entitled to recover whatever items could be positively identified as being his. He would have no chance of recovery, for example, if he knew he had paid for two dozen sink units, but none of the fifty sink units in the contractor's store were clearly marked or set aside as the employer's property. Even marking is not foolproof. The marking has to be temporary in character and it is not unknown for an unscrupulous contractor to affix labels to a set of sanitary fittings until they have been inspected for payment on one job, then to replace them with a different set of labels to obtain payment under a quite different contract.
- *The contractor must provide reasonable proof that the listed items are fully insured for their full value under a policy protecting both employer and contractor in respect of specified perils. The insurance must cover the period from transfer of ownership to the contractor until they are delivered to the site.* This kind of proof can be discharged by giving the employer sight of the policy documents and premium receipts.

A final condition (clause 15.3) states that when the value has been included in an interim payment, the materials and goods become the property of the employer. After that, the contractor must not remove them from the premises except for use on site. The contractor, however, retains liability for damage or loss, the cost of storage, handling and insurance until they are delivered to site.

The gross valuation for the purpose of interim payments is separated into amounts which are and amounts which are not subject to retention.

Clause 30.2A.1 specifies the following amounts which are subject to retention:

- The cumulative value at the appropriate stage. This is the amount noted in appendix 2, alternative A opposite the completed stage.
- The amounts of valuations of changes or of instructions regarding the expenditure of provisional sums which are referable to the

interim payment. For example, if the stages are given as 'block 1, block 2', etc., the amounts must be in respect of changes to Requirements or expenditure of provisional sums in respect of the particular block. The employer must take care that if changes are required, the instruction separates the changes referable to each block. Where provisional sums are included in the Employer's Requirements, they too must be referable to particular stages. If one item is concerned which is involved in each stage, for example the heating installation, the provisional sum should be split into a number of provisional sums. Excluded is the valuation of restoration, replacement or repair of loss or damage and removal and disposal of debris which are to be treated under clauses 22B.3.5 and 22C.4.4.2 as if they were a change in the Employer's Requirements. This somewhat unwieldy provision is intended to ensure that retention is not deducted from the value of this restoration work, because retention will already have been deducted when the work was carried out. This item, therefore, is included in the section below.
- The total value of materials and goods stored off-site which satisfy the provisions of clause 15.2.
- The amount of any adjustment made under clause 38 (use of price adjustment formulae).

Clause 30.2A.4 stipulates the following amounts subject to retention:

- If clause 38 is stated to be the fluctuation provision, and amounts are to be allocated to lift installations, structural steelwork or catering installations, the value of materials and goods delivered to the site for such installations, but not yet incorporated, are to be included. There is a proviso that the goods must not be delivered prematurely and they must be adequately protected. Thus, the contractor will be receiving payment, not only for the stage completed, but also for these goods intended for a future stage. Since the stage amounts are cumulative, however, the payments will be absorbed in future stage payments as they are made.

Clause 30.2A.2 specifies the following amounts not subject to retention:

- Amounts resulting from payments made to or costs incurred by the contractor due to instructions regarding opening up and

testing, royalties, and making good of defects during or after the defects liability period at the employer's cost.
- Amounts noted above in respect of restoration, repair, etc. treated in clauses 22B.3.5 and 22C.4.4.2 as if they were changes in the Employer's Requirements.
- Amounts in respect of loss and/or expense under clause 26 or 34.3 (antiquities).
- Amounts payable to the contractor under fluctuation provisions 36 (contribution, levy and tax fluctuations) and 37 (labour and materials cost and tax fluctuations).

Clause 30.2A.3 sets the amounts to be deducted:

- Appropriate deductions following instructions that defects are not to be made good during or after the defects liability period.
- Amounts allowable under the fluctuation provisions 36 (contribution, levy and tax fluctuations) or 37 (labour and materials costs and tax fluctuations).

10.2.2 Periodic payments – alternative B

If the employer wishes payment to be made by periodic payments, he must complete alternative B of appendix 2. If no period is stated, the period between applications for payment will be one month. Since the contrary is not stated, this is a calendar month. It is open to the employer to allow the inclusion of the value of materials and goods off-site in periodic payments. It is important that this information be included in the Employer's Requirements to allow the contractor to take it into account when formulating his Proposals. Where the value of off-site materials and goods are to be included, such inclusion is subject to the attachment of a list and the satisfaction of the appropriate conditions noted above in connection with off-site values in stage payments.

Once again the gross valuation for the purpose of interim payments is separated into amounts subject or not subject to retention.

Clause 30.2B.1 specifies that the following are subject to retention:

- The total value of work, including design work, properly executed including valuations of changes and expenditure of provisional sums, but excluding the valuation of restoration, etc. under clauses 22B.3.5 and 22C.4.4.2 as noted above under the similar provision for stage payments.

- The total value of materials and goods delivered to site and intended for incorporation in the Works. There is a proviso that the goods must not be delivered prematurely and they must be adequately protected against the weather and other things.
- The total value of materials and goods stored off-site which satisfy the provisions of clause 15.2.
- Any adjustment under clause 38 (use of price adjustment formulae) if applicable.

Clause 30.2B.2 stipulates the following amounts which are not subject to retention:

- Amounts resulting from payments made or costs incurred by the contractor due to instructions regarding opening up and testing, royalties, making good of defects during or after the defects liability period at the employer's cost.
- Amounts noted above in respect of restoration, repair, etc. treated in clauses 22B.3.5 and 22C.4.4.2 as if they were changes in the Employer's Requirements.
- Amounts in respect of loss and/or expense under clause 26 or 34.3 (antiquities).
- Amounts payable to the contractor under the fluctuation provisions 36 (contribution, levy and tax fluctuations) and 37 (labour and materials cost and tax fluctuations).

Clause 30.2B.3 sets out the amounts to be deducted. These amounts are not subject to retention:

- Amounts allowable under the fluctuation provisions 36 and 37.
- Appropriate deductions following instructions that defects are not to be made good during or after the defects liability period.

10.3 Applications

It is for the contractor to apply to the employer for payment. To that extent he is in the driving seat. The scheme of the contract is that the contractor carries out the valuation described in sections 10.2.1 or 10.2.2, as appropriate. After that, it is for the employer to pay unless he disputes the amount. Clause 30.3.6 provides that the employer has 14 days from the date of receipt of each interim application in which to pay the contractor. Applications are dealt with in clause 30.3.1.

If it is stated in appendix 2 that payment is to be made by stages, the contractor must make application for interim payment on completion of each stage. He may make one further interim application which is to be after the end of the defects liability period or on the issue of the notice of completion of making good defects, whichever is later. This is to deal with the release of the second half of the retention. Something which may cause difficulty is that 'completion' of a stage is not defined. It is probably sensible to consider 'completion' very much the same kind of condition as 'practical completion' of the project as a whole, but applied to the particular stage described in the schedule. When considering the meaning of 'completion' in clause 27.4.4 of JCT 80 in *Emson Eastern Ltd* v. *E.M.E. Developments* (1991), the judge said:

> 'In my opinion, there is no room for "completion" as distinct from "practical completion". Because a building can seldom, if ever, be built precisely as required by drawings and specification, the contract realistically refers to "practical completion" and not "completion", but they mean the same.'

The only problem which is likely to arise with regard to the last interim application is that the employer may be dilatory in issuing the notice of completion of making good defects. In no sense is the notice intended to be a certificate such as the architect issues under JCT 98 and it is notable that the word 'certificate' has been avoided. Nevertheless, it is thought that the right to make application depends on it. A contractor faced with the employer's failure to issue should make a formal request for such notice followed by a requirement for adjudication on the matter. It should be capable of very quick resolution. It is not something which entitles the contractor to determine his employment under clause 28.2.1 (see Chapter 12), because the contractor's right to payment does not arise until an application under clause 30.3.1 has been made.

If it is stated in appendix 2 that periodic payments apply, the contractor must make application at whatever periods are stated in appendix 2 alternative B. Note that this is a duty. The contractor is obliged to make application as prescribed. For example, the contractor may not forego making application because he wants to keep his income low in a particular period for tax purposes. The application must be made and then the employer has a good idea, from month to month, just how the finance is working. Were it not for this duty placed on a contractor, he could, for his own purposes, virtually make application to suit his own cash requirements.

There is no particular starting date set out for the applications, just that they must be made at the stipulated periods apart. It is suggested that, if one month is the period in alternative B, the first application should be made one month after the contractor takes possession of the site. Applications must continue to be made until the end of the period in which practical completion occurs. For example, if the date for possession is 3 September 1999 and the date of practical completion is 2 May 2000, the first application would be made on 3 October 1999 and the last regular application would be made on 3 May 2000. After practical completion, applications must be made not at regular intervals but when further amounts are due, particularly after the end of the defects liability period or on the issue of the notice of completion of making good defects, whichever is later. There is a proviso that the employer may not be required to make payment within one calendar month (whatever the period for interim applications may be) of having made a previous interim payment.

Clause 30.3.2 makes the important proviso that each application must be accompanied by whatever details the employer has set out in his Requirements. This applies whether the payments are to be by stages or periodic. The employer will certainly wish to check the contractor's application before payment and he will usually do this through the medium of his agent, but if the agent is an architect or an engineer, the employer may employ a quantity surveyor for the purpose depending on the size and complexity of the project. It is essential, therefore, that the details requested in the Employer's Requirements are such as will assist the employer or his agent in checking the application. In the absence of any precise specification of such details in the Employer's Requirements, it is considered that the contractor would be entitled to submit his application as a lump sum which would be almost impossible to check. Applications for stage payments are easiest to check even though the contract provides that the actual figure applied for may be somewhat more (or less) than set out in the stages (see section 10.2.1). Correspondingly greater thought must be given to the calculation of the amounts of stage payments before the contract is executed.

Under the provisions of clause 30.3.3 and in order to comply with the Housing Grants, Construction and Regeneration Act 1996, the employer must give the contractor written notice not later than five days after receipt of the contractor's application. The notice must specify the amount of payment the employer proposes to make, to what it relates and the basis of calculation. It is probably of little consequence if the employer forgets to give this notice provided he pays in full by the final date for payment. If he fails to pay by the due date, failure to give this notice might well have serious consequences on his right to withhold payment unless his notice under clause 30.3.4 (see below) falls into the same time frame. Although this would be implied from the operation of the Act, it is expressly stated in clause 30.3.5.

If the employer fails to pay any amount due to the contractor by the final date for interim payment (i.e. 14 days from the date of receipt by the employer of the contractor's application for payment), clause 30.3.7 requires the employer to pay simple interest on the outstanding amount at the rate of 5% above Bank of England current base rate. It is to be treated as a debt. In other words, the contractor can bring an action for recovery. The clause makes clear that payment of the interest is not to be taken as a waiver of the contractor's other rights, namely his right to proper payment of the principal owing at the right time, or his right to suspend performance of his obligations, or his right to determine his employment. It is another contractual remedy for late payment and it can be exercised whether or not the contractor also opts to suspend or determine. An employer in financial difficulties is not to treat this provision as a kind of extended loan arrangement albeit with high interest rates.

10.4 *Employer's right to withhold payment or to deduct*

The employer's contractual right of withholding, or deduction from, money due to the contractor is contained in clause 30.3.4. The employer may exercise any right under the contract against any amount due to the contractor even if retention over which the employer is trustee is included in the amount (clause 30.4.3). If the employer considers that he is entitled to withhold or deduct an amount from a payment due to the contractor, he must give a written notice to the contractor not later than five days before the final date for payment. The notice must state the grounds for withholding payment and the amount of money withheld on each ground. The information must be detailed enough to allow the contractor to understand the reason why he is not receiving the amount withheld. Of course, the contractor may seek immediate adjudication under clause 39A (see Chapter 13). The contract gives the employer the right of deduction under the following clauses:

4.1.2 The cost of employing others following non-compliance with an employer's instruction.

16.2 Appropriate deduction after instruction not to make good defects.
16.3 Appropriate deduction after instruction not to make good defects.
21.1.3 The cost of insurance premiums following the contractor's failure to insure.
21.2.3 The cost of insurance premiums following the contractor's failure to insure.
22A.2 The cost of insurance premiums following the contractor's failure to insure.
24.2 Liquidated damages.
31.7 Statutory tax deduction scheme where the employer is a 'contractor'.

In addition to the employer's contractual right of deduction, he also has the common law right of set-off which can only be excluded by an express clause to that effect: *Gilbert Ash (Northern) Ltd* v. *Modern Engineering (Bristol) Ltd* (1973). Although at one time it was rare for an employer to exercise this right, it is becoming more common, certainly under traditional contract forms, but the employer must have a genuine reason for set-off if he is to succeed (*C.M. Pillings & Co Ltd* v. *Kent Investments Ltd* (1985) and *R.M. Douglas Construction Ltd* v. *Bass Leisure Ltd* (1991)) and he is obliged to comply with clause 30.3.4.

Clause 30.3.4 is important from the employer's point of view. It provides a deceptively simple machinery for the employer to withhold payment and it applies only to interim payments. It is triggered by receipt of the contractor's application for an interim payment. It is not good enough for the employer simply to say in his notice that some of the work included in the application is defective. He must give a reasonable description and quantification of the sum withheld so that the contractor can consider his position.

Whether the amount stated in the contractor's application or the payment made by the employer is in accordance with the contract can only be decided by looking at the contract provisions in any particular circumstance. The onus appears to be on the employer to show in his notice that the amount claimed is not in accordance. The likely reason why an amount is not in accordance with the contract will be because it does not comply with the payment provision in clauses 30 to 30.2B inclusive. It could be, for example, that the employer disputes that a stage has been completed, or that materials delivered to site are too early. Experience suggests that the major reason given by the employer for withholding payment will be because he contends that there are defects in the work. In such a case, he would have to demonstrate that the workmanship and/or materials did not comply with clause 8.1. It would be helpful to his contention if, before the application, the employer has used powers under clause 8.4 in respect of the alleged defects. In the past, many employers simply deduct round figures vaguely referenced to 'defects in the work'. In our view, an employer behaving in this manner is asking for, and deserves, a notice of determination. The intention of these clauses is clear. It gives the contractor a leading role in obtaining payment, but it provides the employer with a remedy within strict procedures if the contractor tries to get more than his entitlement. Both parties should treat the whole of clause 30.3 with great respect.

10.5 Retention

The rules for ascertainment of retention are set out in clause 30.4 and they deserve careful study. The provisions are straightforward. The employer is entitled to retain a percentage of the total value of work, materials and goods included under either clause 30.2A.1 (where stage payments apply) or clause 30.2B.1 (where periodic payments apply). The percentage is 5% unless the parties have inserted a lower rate in appendix 1. There is no provision for the insertion of a higher rate, but if the employer so desired, and the contractor agreed, there is no reason why the clause should not be amended slightly to allow the insertion of a higher rate in the appendix. As the contract figure increases, the amount retained becomes progressively greater and it is a factor which the contractor has to take in account when calculating his tender. This is probably one of the reasons why a footnote advises that if the contract sum is £500,000 or over, the retention percentage should not be more than 3%. That still amounts to £15,000, a substantial sum. The retention is to be released in two equal parts: part one after the work has reached practical completion and part two after the notice of making good defects has been issued. If partial possession has been taken by the employer under clause 17, the retention is released in proportion to the value of the 'relevant part' taken into possession, and subsequently a further and equal amount when notice of making good defects of that part has been issued.

Under clause 30.4.2, the retention is said to be subject to certain rules. Clause 30.4.2.1 states that the employer's interest in the retention is fiduciary, but without obligation to invest. This clause acknowledges that the money thus retained is in reality the contractor's money which the employer is simply holding in trust for the contractor. On the face of it, this is good news for the contractor, because it means that if the employer should become insolvent, the retention money should be safe and protected away from the clutches of creditors. It is well established, however, that in the event of an insolvency, such a trust fund is effective only if it is readily identifiable as belonging to the contractor. Clause 30.4.2.2 therefore provides that if the contractor so requests, the employer must put aside into a separate bank account the amount of retention deducted at the date of each interim payment. The account must be identified as trust money held under the contract provisions and the employer must give the contractor written notice to that effect. The employer is entitled to the full beneficial interest and has no duty to account for it to the contractor.

The contractor's right to have the retention put into a separate account as stipulated by this clause has been supported by the courts who have been prepared to issue mandatory injunctions for the purpose: *Rayack Construction Ltd* v. *Lampeter Meat Co Ltd* (1979). More recently, the Court of Appeal has decided that even in the absence of an express obligation to place the retention in a separate bank account, the employer is still obliged to do so: *Wates Construction (London) Ltd* v. *Franthom Property Ltd* (1991). In that case, which concerned a very similar retention clause to that in WCD 98, the clause obliging the employer to put the retention in a separate bank account was deleted. The court said:

> '...clause 30.5.1 creates a clear trust in favour of the contractors and subcontractors of the retention fund of which the employer is a trustee. The employer would be in breach of his trust if he hazarded the fund by using it in his business and it is his first duty to safeguard the fund in the interests of the beneficiaries...'

The court had this to say about the deletion:

> 'Firstly, it seems to me that there is no ambiguity about the part of the agreement which remains. The words of clause 30.5.1 under which the trust is created are quite clear. Secondly, the fact of deletion in the present case is of no assistance because the parties, in agreeing to the deletion of clause 30.5.3, may well have had different reasons for doing so and it is not possible to draw from the deletion of that clause a settled intention of the parties common to each of them that the ordinary incidence of the duties of

trustees clearly created by clause 30.5.1 were to be modified or indeed removed. It may have been thought by one of the parties to have been unnecessary to have included clause 30.5.3. It may have been that one of them thought that the employer should have been liable to account for any interest to the contractor if the retention fund was placed in a separate account. But there may be various reasons, which it is not possible to set out in full, why the clause was deleted and it is quite impossible to draw any clear inference from the fact of deletion. I therefore would reject an argument based upon the fact of deletion and can see no ambiguity upon which reference to that deleted clause could assist.'

It is established that the contractor does not have to make a request to have the retention put into a separate bank account every time an interim payment is made. Indeed, it seems that the employer has the obligation whether or not there is a clause to that effect and whether or not the contractor so requests: *Concorde Construction Co Ltd* v. *Colgan Co Ltd* (1984). The employer cannot rely on his right to deduct liquidated damages to extinguish the retention fund and so leave him with no obligation in respect of a separate account where WCD 98 is used: *J.F. Finnegan Ltd* v. *Ford Seller Morris Developments Ltd* (1991). The Court there held that the employer's own notice of non-completion did not have the same binding effect as did an architect's certificate issued in similar circumstances. Determination of the contractor's employment under the contract has no effect on his right to require the retention fund to be set aside. It appears, however, that there is no trust established until the separate fund has been set aside and if the employer's insolvency predates the setting aside, the contractor will be in no better position in respect of the retention than would any other ordinary creditor: *MacJordan Construction Ltd* v. *Brookmount Erostin Ltd (In Administrative Receivership)* (1991).

A footnote to clause 30.4.2.2 states that it should be deleted if the employer is a local authority. In the light of the case law noted above, it seems that deletion would be ineffective. If the employer wishes to avoid setting up a separate account, it may be that an appropriately worded clause in place of clause 30.4.2.2 would suffice. In that case, and despite clause 30.4.2.1 designating the fund as a trust, it seems that a trust would never arise. The position is now unsatisfactory if the employer avoids setting up a separate account and contractors should request a separate account and ensure that it is set up. For safety, the separate bank account should state:

- the name of the project; *and*
- that it is a trust fund; *and*
- that the employer is trustee; *and*
- that the contractor is the beneficiary.

The bank manager should be informed by letter from the employer, with a copy to the contractor, that it is a trust fund under clause 30.4.2 of the contract. If this is not done and appropriate evidence supplied to the contractor within a reasonable time, the contractor has no real alternative but to seek a mandatory injunction to enforce his rights under contract.

We express no opinion about whether the clear words of the contract are sufficient to avoid what appears to be the employer's statutory duty as trustee to account for the interest to the contractor. The duties of a trustee are set out in the Trustee Act 1925 and the Trustee Investments Act 1961. No doubt the question will come before the courts in due course. By virtue of clause 30.4.3, the employer is entitled to exercise any right of deduction from money due or to become due to the contractor even if retention is part of that money.

10.6 Final payment

The process is triggered by the written statement of practical completion under clause 16.1. Within the three months following, clause 30.5.1 stipulates that the contractor must submit the final account and final statement for the employer's agreement. The contractor must supply the employer with 'such supporting documents as the Employer may reasonably require'. The contractor must supply such documents within the three month period, therefore it is essential that the employer informs the contractor of his requirements in good time. If the employer neglects to inform the contractor, the contractor should take the initiative by requesting details. In practice, the employer may argue that he cannot possible state what supporting documents will be required until the contractor's final account is received. Although that is certainly true so far as detail is concerned, there is nothing to prevent the employer from letting the contractor know the kind of documents required. There is nothing in this clause to suggest that the employer must state in the Employer's Requirements the documents required. Were that intended, an express statement would have been included such as is contained in clause 30.3.2. The requirement for documents must be reasonable.

Clause 30.5.2 provides that the contract sum is to be adjusted in accordance with the clauses of the contract. Presumably for the avoidance of doubt, it has been thought necessary to state that this includes agreements made under clause 12.4.1 and matters in connection with the price statement under clause 12.4.2. The final account must show the way in which the contract sum is adjusted. Thus adjustments are set out in clause 30.5.3.

Five categories are to be deducted:

(1) All provisional sums in the Employer's Requirements
(2) Any deductions resulting from clause 2.3.2 (boundary/ Employer's Requirements discrepancies) and clause 6.3 (change in statutory requirements after base date)
(3) The amount of any valuation of omissions resulting from a change together with amounts resulting from consequential changes in conditions of work
(4) Any deductions under clauses 16.2 or 16.3 (appropriate deductions for defects not made good) or the fluctuation clauses 36, 37 or 38
(5) Any other amounts deductible under the provisions of the contract.

Nine categories are to be added:

(1) Any additions resulting from clause 2.3.2 and clause 6.3
(2) Amounts payable under clauses 8.3 (inspections and testing), 9.2 (royalty payments), 16.2 or 16.3 (payments for making good)
(3) Amounts of valuations of changes together with amounts resulting from consequential changes in conditions of work
(4) Amounts of valuation of work in accordance with instructions on the expenditure of provisional sums
(5) Amounts ascertained under clauses 26 (direct loss and/or expense) and 34.3 (direct loss and/or expense following the discovery of antiquities)
(6) Amounts paid under clauses 22B.2 or 22C.3 after the employer's failure to maintain insurance
(7) Amounts payable under the fluctuation clauses 36, 37 or 38
(8) Amounts paid under clause 12.4.2 alternative A7 acceptance of the price statement in lieu of ascertainment under clause 26.1
(9) Any other amounts to be added under the provisions of the contract.

The final statement must set out the amount which results after the contract sum has been adjusted and also the total amount which has already been paid to the contractor. The difference between the two amounts is to be shown as a balance payable either by the employer or by the contractor as appropriate. The statement must say to what the balance relates and the basis of calculation.

Clause 30.5.5 sets out a further timescale. The employer has one month to dispute the final account or final statement as submitted by the contractor. The period is measured from :

- the end of the defects liability period; *or*
- the day named in the notice of completion of making good defects; *or*
- the submission of the final account and final statement by the contractor

whichever is latest.

If the employer does not dispute it during the prescribed period, the final account and final statement are conclusive regarding the balance due to either the contractor or to the employer as appropriate. If the employer does dispute it or disputes part of it, the final account and final statement are conclusive only in respect of the part not disputed. It appears to follow that if the whole of the account and statement are disputed, none of it becomes conclusive.

If the contractor fails to submit his final account and final statement within the three month period, the employer can take matters into his own hands under clauses 30.5.6 to .7. The employer may give the contractor two months written notice that he intends to prepare the final account and final statement himself (or have it prepared by another person on his behalf). If the contractor does not respond, the employer may proceed himself after the expiry of the two months. Essentially, the employer must go through the same process as would the contractor, deducting and adding amounts to adjust the contract sum in accordance with clause 30.5.3. The contract recognises that the employer is restricted to using only the information in his possession. The employer's final statement must set out the amounts in the employer's final account and the total amount already paid to the contractor and must indicate the difference as payable by the employer or by the contractor as appropriate. Without the contractor's input, it is very likely that the final balance will be defective. If the employer has done his best despite the lack of information, any deficiencies will be the result of the contractor's failure to act as required by the contract. He is given one month to dispute the employer's final account and final statement by clause 30.5.8. The period is measured from:

- the end of the defects liability period; *or*
- the day named in the notice of completion of making good defects; *or*
- the submission of the final account and final statement by the employer

whichever is the latest.

If the contractor does not dispute it within the period, the employer's final account and final statement are conclusive regarding the amount due to either the contractor or to the employer as appropriate. Our comments above regarding dispute or partial dispute of the employer's final account and employer's final statement are also applicable to this situation.

The contract is silent about the procedure to be followed in such cases. Presumably, it is for the employer and the contractor to try to settle the dispute by agreeing an amount to replace the amount disputed. The dispute is something which could be referred to adjudication under clause 39A or to arbitration or be determined by legal proceedings as appropriate under clauses 39B or 39C respectively. The problem is that total conclusivity is only achievable under clause 30.5.5 if the employer disputes nothing. Although the contract seems to recognise the concept of partial conclusivity in clauses 30.5.5 and 30.5.8, other clauses do not seem to accept anything other than total conclusivity. A number of matters hinge on conclusiveness under clauses 30.5.5 and 30.5.8. The timing of notices and, in particular, payment under clauses 30.6.1 and 30.6.2 and the conclusivity of the final statement in respect of various matters under clause 30.8.1 are effectively now dependent on the final statement not being disputed. Once a dispute is registered, there is no defined method of gaining the conclusiveness. It is disturbing to think that the contract may have the seeds of its own frustration in these clauses.

If the final statement is prevented from being conclusive in respect of the matters set out in clause 30.8.1, it may be irritating, but in most cases it will not be disastrous. Prevention of the final payment would be very serious. There is no immediately obvious answer. In most cases, no doubt the parties will agree a final account in due course and they will observe the final payment provision from that point. If the employer chooses to be obstructive, however, the contract does not provide any immediately obvious solution, at least so far as we can see. We put forward the suggestion that in order to give the contract business efficacy, a term would have to be implied to the effect that in the event of an initial dispute the final account and final statement would be conclusive regarding the balance on the date that agreement was reached and recorded by the employer and the contractor in respect of the part or parts disputed or, failing agreement, on the date of a decision of an adjudicator, award of an arbitrator or judgment of a court as the case may be. The last thing needed in a contract form of such length and complexity is uncertainty about a significant term and JCT should deal with this point at the earliest opportunity. Meanwhile, users should consider amending these provisions.

Clause 30.6.1 stipulates that the employer must give written notice to the contractor specifying the amount of payment to be made. The notice must be given not later than five days after the final statement or the employer's final statement becomes conclusive about the balance due between the employer and the contractor. This notice is for the same purpose in respect of the final payment as the notice given under clause 30.3.3 in respect of applications for interim payment.

Clause 30.6.2 provides that 28 days after the final statement or the employer's final statement becomes conclusive is the final date for payment of the balance in the final statement by one or the other of the employer or the contractor as applicable. If the employer intends to withhold or deduct payment from the final payment, he must give a written notice to the contractor not later than five days before the final date for payment. The notice must state the amount to be withheld, the grounds for withholding and the amount to be withheld in respect of each ground. This provision echoes clause 30.3.4 in regard to applications for interim payment. If the employer fails to give the notices required under clauses 30.6.1 or 30.6.2, he is obliged to pay the balance, if any, stated as due in the final statement or the employer's final statement without any deduction (clause 30.6.3).

If the employer or the contractor fails to pay the balance or the appropriate part (after allowable deductions) to the other by the final date for payment, the party owing the money must pay simple interest at the rate of 5% over the current base rate of the Bank of England as well as the amount not properly paid. Payment of interest is not to be construed by the paying party as a waiver of the receiving party's rights to full payment (clause 30.6.4).

Clause 30.8.1 states that when the final statement or the employer's final statement has become conclusive (see our comments above) regarding the balance due between the parties, it is conclusive evidence in any proceedings arising out of or in connection with the contract whether by arbitration or litigation that:

- where it is expressly stated in the Employer's Requirements or in an instruction that the particular quality of materials or the standards of workmanship are to be to the approval of the employer, the quality and standards are to his reasonable satisfaction. This, therefore, applies only where the employer has expressly stated in his Requirements that quality or standards are to be to his satisfaction. For example, he may want to retain control over the floor finishes or the standard of brickwork. In this instance, the final account and final statement would be conclusive that the floor finishes or brickwork were to the employer's reasonable satisfaction irrespective of whether the employer had actually taken the trouble to examine them. The employer should, therefore, take particular care not to include such clauses in his Requirements unless they are absolutely necessary and unless the employer or his agent are certain to check them on site. The clause as originally drafted was tighter than the equivalent clause in JCT 80 and it is possible that it would have escaped the effects of the decision in *Crown Estates Commissioners* v. *John Mowlem & Co* (1994). The drafting of this clause has been tightened still further following *Crown Estates* to make clear that the final statement is not conclusive about the employer's reasonable satisfaction except to the limited extent expressly set out in the Employer's Requirements. It also now states that the final statement is not conclusive that any materials or workmanship comply with the contract, even the ones about which the employer is said to be reasonably satisfied. The precise result of that remains to be seen.
- all extensions of time due under clause 25, and no more, have been given.
- reimbursement of loss and/or expense under clause 26 is in final settlement of all claims, whether for breach of contract, duty of care, statutory duty or otherwise, which the contractor may have and which arise out of any of the 'matters' in clause 26.2.

Proceedings are expressly stated to include adjudication under article 5, arbitration under article 6A or legal proceedings under article 6B. The reference to extensions of time and loss and/or expense were added because, in some instances, contractors were making very late claims in contract or in tort, long after the employer thought that he had discharged his final payment.

The conclusive effect of the final statement is subject to four exceptions. It is not conclusive:

(1) if there is any fraud. Fraud is the tort of deceit and may be a misrepresentation made knowingly or without belief in its truth, or recklessly careless whether it is true or false: *Derry* v. *Peek* (1889).

(2) if adjudication, arbitration or other proceedings have previously been commenced (clause 30.8.2). In this case, they are conclusive in respect of the specified topics after the proceedings have been concluded, but subject to the terms of any award or judgment or settlement or after 12 months from the issue of the final account and final statement if neither party has taken any further step in the proceedings, but subject to the terms of any partial settlement. The latter part of this exception is clearly intended to thwart a party who may serve notice of arbitration as a holding measure to prevent the final account and final statement becoming conclusive, but who concludes that it is not worthwhile taking the matter further. If it was not for this provision, service of the one notice would be sufficient to prevent the final account and final statement from ever becoming conclusive about the specified matters.

(3) if adjudication, arbitration or other proceedings have been commenced by either party within 28 days after the final account and final statement or employer's final account and employer's final statement would otherwise become conclusive, they are conclusive as specified except in respect of matters to which the proceedings relate (clause 30.8.3).

(4) if the adjudicator gives his decision on a dispute after the date of submission of the final account or final statement or the employer's final account or employer's final statement, when the employer or the contractor have 28 days, from the date on which the decision was given, to commence arbitration or legal proceedings to finally determine the dispute. This provision is to protect the right of either party to require arbitration or legal proceedings after an adjudicator's decision which, taking account of agreed extension of the statutory period, may not be given until more than 28 days after the final account and final statement become conclusive.

A very important clause (30.9) provides that no payment made by the employer is of itself conclusive evidence that any design, works, materials or goods are in accordance with the contract except as noted above. This prevents the contractor contending that work carried out must be properly executed because he included it in an interim application which the employer paid without query. It is thought that, since the applications are cumulative (see clauses 30.2A.1.1 and 30.2B.1.1), the employer may rectify any overpayment of this sort by simple notice to the contractor, in accordance with clause 30.3.4, at least five days before the final date for the next payment.

10.7 *Advance payment*

The contract provides in clause 30.1.1.2 for advance payment by the employer to the contractor. It is to be deleted, together with clause 30.1.2.2, if the employer is a local authority. The idea is that the employer pays the contractor a lump sum, usually at the start of the project, and the contractor repays it over a period of months. It is for the employer to decide whether or not to operate this clause. If the employer decides to make advance payment, appendix 1 must be completed accordingly by inserting the amount to be paid and the times and amounts for repayment. The appendix must also state whether the employer wants a bond. Normally, we would expect that if the employer is prepared to make advance payment, he will certainly require a bond. The bond is to be provided by a surety approved by the employer and the employer need not make payment until the bond has been provided. The terms of the bonds are to be those agreed by the British Bankers' Association and the JCT unless the employer requires other terms. If other terms are required, the employer must have given the contractor a copy before the contract is executed (sixth recital). The date on which the employer must make the advance payment is also to be stated in appendix 1. A sample of the bond on agreed terms is bound into the contract as annex 1 to appendix 1. There will be many instances where the contractor will welcome an advance payment to enable him to fund the start of the project. In turn, the employer will doubtless expect to secure a price advantage. From the contractor's point of view, the repayment amounts should be carefully calculated so that they are amply covered by the expected interim payments.

10.8 *Changes*

Changes are dealt with in clause 12, which is very similar to the variation clause (clause 13) of JCT 98. The term 'change' is defined in clause 12.1. It means 'a change in the Employer's Requirements which makes necessary the alteration or modification of the design, quality or quantity of the Works, otherwise than such as may be reasonably necessary for the purposes of rectification pursuant to clause 8.4 . . .'. (It should be noted in passing that the reference to rectification must be an error, because rectification has been removed from clause 8.4 – see our comments in section 5.3). This broad definition is stated to include four distinct categories:

(1) Addition, omission or substitution of any work.
(2) Alteration of the kind or standard of materials or goods.
(3) Removal of work carried out, materials or goods brought on site by the contractor for the Works and which are not defective.
(4) The imposition, addition, alteration or omission of any obligations or restrictions imposed by the employer in his Requirements or by a change in regard to access or use of the site, limitation of working space or hours or the carrying out of the work in a particular order. The employer may impose such obligations even though none were in the Employer's Requirements. The contractor has the right of reasonable objection to this kind of change under clause 4.1 (see section 5.3.1).

Clause 12.2.1 empowers the employer to issue instructions to produce a change in his Requirements. This is subject to the proviso that he may not produce a change which requires an alteration in design unless the contractor consents. This point is discussed in section 3.2. Where the contractor is also acting as planning supervisor, he must notify the employer whether he has any objection under section 14 of the CDM Regulations and, if he has, the employer must amend the instruction and the contractor is not obliged to comply until the amendment has been made (clause 12.2.2). No change will vitiate the contract. In one sense this is superfluous wording because it is clear that the exercise of a power which is provided for in a contract by agreement of the parties cannot bring that contract to an end. However, the provision must be considered in relation to the first recital which briefly describes the Works. A change, or a number of changes which resulted in the description in the first recital becoming inaccurate would certainly be beyond the power of the employer. It is thought that the limit to the employer's power would occur before that stage was reached. A change must not only be within the meaning given in clause 12.1, it must also bear some relation to the Works as described in the first recital.

Clause 12.3 provides that the employer may issue instructions in relation to provisional sums if they are in the Employer's Requirements. He has no power to issue any instructions if the provisional sums are included in the Contractor's Proposals.

The valuation rules closely echo the equivalent rules in JCT 98. Clause 12.4.1 sets the scene by stipulating that the valuation of changes and provisional sum work is to be valued by reference to clause 12.4.2, alternative A. The employer and the contractor can agree otherwise. The contractor is the driving force in this contract and if he opts not to use alternative A, where there is no other agreement and where supplementary provision S6 does not apply, the valuation is to be carried out under clause 12.4.2, alternative B. Valuations, however they are made, must include allowance for addition and, perhaps strangely, for omission of relevant design work. It is essential that the Contract Sum Analysis sets out a method of valuing design work. It must be rare for there to be an omission of design work because it is normally done so far in advance of the construction.

Under option A and to add complication to the already complex valuation provisions, the text is divided not into clauses but into paragraphs. Paragraph A1 provides that the contractor has 21 days from the date he receives an instruction, or from the date on which he has received sufficient information, to submit a price statement to the employer. The statement must show his price for the work which must be based on the clause 12.5 valuation rules. The contractor may also, entirely at his option, separately submit a figure for related loss and/or expense and any extension of time he requires. The employer under paragraph A2 has 21 days to write to the contractor either accepting the price statement or notifying the contractor that some or all of it is not accepted. If there is no response, the employer is deemed not to have accepted and the contractor may refer the statement as a dispute or difference to the adjudicator under clause 39A. If there is a total or a part acceptance, the whole or the appropriate part of the price statement must be added or deducted from the contract sum (paragraph A3).

Paragraph A4 stipulates that if the employer does not accept, he must give reasons in the same detail as contained in the contractor's statement and he must enclose an amended statement which he is prepared to accept. The contractor has 14 days from receipt of the employer's price statement in which to decide whether he accepts or not. Insofar as he accepts, the sum is added to or deducted from the contract sum as before. If the contractor does not respond within 14 days, or if he does not accept any part of it, the employer or the contractor may refer the contractor's price statement and the amended price statement as a dispute or difference to the adjudicator under clause 39A. In the event that neither party has accepted nor made any reference to the adjudicator, paragraph A6 states that alternative B applies. The employer may either accept or reject the contractor's separate submissions dealing with loss and/or expense and extensions of time. If he rejects them or if he simply fails to respond within 21 days of receipt, paragraph A7 prescribes that clauses 25 and 26 will apply, the contractor's submissions being effectively consigned to the wastebasket.

If alternative B is to apply, the valuation must be carried out under clauses 12.5.1 to 12.5.6. The text numbering reverts to clauses under alternative B. The reason for that is that alternative B is essentially the valuation rules which existed in CD 81 before the concept of the contractor's price statement was introduced. The contract does not stipulate who is to carry out the valuation, but in view of the contractor's responsibility to make applications for payment under clause 30.3.1, it is clear that the contractor must also carry out valuations. This appears to give him considerable power, but valuations are to be carried out objectively. There is no question of opinion and if the employer disputes the sum he may use his powers under clause 30.3.4 (see section 10.4) which may cause the contractor to seek immediate adjudication under clause 39A. If the supplementary provisions are in operation, different procedures will apply (see section 10.9). There is no particular procedure in the rules for the valuation of design work which is or which becomes abortive. This is a point which might warrant some amendment of this clause or the contractor may lose substantial sums if faced with an employer who constantly changes his mind.

The principal document to be consulted when applying the rules is the Contract Sum Analysis. If the supplementary provisions apply and the employer has opted to describe his Requirements in terms of bills of quantities in accordance with S5, the principal document becomes the bills of quantities. It is difficult to imagine an employer normally using this approach.

Omissions must be valued in accordance with the values in the Contract Sum Analysis. No adjustments are needed (clause 12.5.2).

Clause 12.5.1 provides that additional or substituted work may be valued in one of three ways:

(1) Work of similar character to the work in the Contract Sum Analysis: the valuation must be consistent with the values in the Contract Sum Analysis.
(2) Work of similar character to the work in the Contract Sum Analysis, but with changes in the conditions under which the work is carried out or changes in the quantity of work: the valuation must be consistent with the values in the Contract Sum Analysis with due allowance made for the changes.
(3) Where there is no work of a similar character to the work in the Contract Sum Analysis: a fair valuation must be made.

If the proper basis for fair valuation is considered to be daywork, the valuation must comprise the prime cost of the work together with percentage additions on each section of the prime cost at rates set out by the contractor in the Contract Sum Analysis. It is vital that the contractor includes such rates whether specifically requested or not. The contract spells out what is acceptable in terms of daywork. It is to be calculated in accordance with the 'Definition of Prime Cost of Daywork carried out under a Building Contract', issued by the Royal Institution of Chartered Surveyors and the Building Employers' Confederation (now renamed the Construction Confederation). Alternatively, if the work is within the province of any specialist trade and there is a published agreement between the RICS and the appropriate employers' body, the prime cost is to be calculated in accordance with the definition in such agreement current at the base date. Where the contractor considers that the appropriate basis of valuation is daywork, he must submit what the contract persists in calling 'vouchers', but what everyone in construction knows are 'daywork sheets'. They must reach the employer not later than the end of the week following the week in which the work was carried out and they must contain the names of workmen and the plant and materials employed together with the time spent doing the work. There is no contractual requirement that the vouchers should be signed by the employer or his agent, but such a signature is usual, signifying the employer's agreement that the vouchers are factually correct. It is not subsequently open to the employer to refuse payment on the grounds of errors in times, personnel, etc. (*Clusky* v. *Chamberlain* (1995); *Inserco* v. *Honeywell* (1996)), but he may argue that daywork is an inappropriate basis.

Clause 12.5.3 provides that valuations must include an allowance for addition or reduction of site facilities, site administration and temporary works – the equivalent to what JCT 98 would term 'preliminaries'.

It may happen that as a result of the contractor carrying out an instruction requiring a change or the expenditure of a provisional sum, the conditions under which other work is carried out are altered substantially. If so, the other work must be treated as if the alteration results from an instruction and it must be valued accordingly. This provision echoes a similar term in JCT 98.

There is an overriding proviso that no allowance must be made in any valuation for the effect on regular progress of the Works or for any direct loss and/or expense for which the contractor would be reimbursed by any other provision of the contract. It is clear that a distinction is to be drawn between this clause and clause 26, under which the contractor will usually recover direct loss and/or expense. It is possible for the contractor to recover direct loss and/or expense under clause 12.5 if he can show that no other clause will cover the particular point.

Clause 12.6 provides that valuations carried out in accordance with clause 12.5 are to be added to or deducted from the contract sum. This is a fairly pedantic, but necessary, provision to link with clause 3 so that the amount of such valuations can be included in the next interim payment after the valuation has been carried out.

10.9 Valuation of changes under the supplementary provisions

This is supplementary provision S6. Where supplementary provisions apply, this one modifies clauses 12, 25 and 26 – changes, extensions of time and loss and/or expense respectively. The idea behind the provision is that all the effects of a change instruction can be assessed and dealt with before or at the same time as the instruction is carried out. It has great similarities to the valuation of instructions under the Association of Consultant Architects' Form of Building Agreement (ACA 2). It imposes a strict discipline on both parties, but particularly on the contractor who pays a severe penalty if he forgets to operate the provision correctly. Contractors should be vigilant to check whether these provisions apply. It is our experience that both employer's agents and contractors frequently deal with the whole contract through to practical completion without realising that there are supplementary provisions. In some ways the introduction of the contractor's price statement into clause 12 as alternative A duplicates this procedure. However, it is clear that if the supplementary provisions apply, they take precedence.

The procedure is triggered by the issue of an instruction by the employer under clause 12.2 requiring a change. If the contractor or the employer decides that the instruction:

- will have to be valued under clause 12.4; *or*
- will result in an extension of time; *or*
- will result in the ascertainment of loss and/or expense

the contractor *must* submit to the employer a set of estimates. He must do this before he carries out the instruction and no later than 14 days after the date of the instruction. The parties may agree a different time limit and, if they cannot agree, the period is to be 'reasonable in all the circumstances'. The provision anticipates a situation arising where the instruction is of such complexity that 14 days is not satisfactory. It is not thought that a particular problem will occur as a result of the contract referring to either contractor's or employer's opinion. In practice, if the contractor is not of the opinion that the instruction will have one or more of the effects noted, it is highly unlikely that the employer will attempt to overrule the contractor to his own disadvantage.

The contractor need not provide the estimates if the employer states in writing, at the time he issues the instruction or within 14 days after that he does not require any estimates or if the contractor puts forward a reasonable objection to any or all of the estimates. The contractor must act within ten days of receiving the instruction and the objection may be from the contractor or from any of his subcontractors. A reasonable objection could be that the instruction is such that the delay effect on the completion date cannot be calculated until the instruction has been completed. In any event, such objection may be referred to the adjudicator for a decision.

The estimates required from the contractor are as follows and they replace valuation and ascertainment of the instruction under clause 12.4 and 26 respectively:

- The value of the instruction. This is, to all intents and purposes, a quotation and it must be supported by full calculations referenced to the Contract Sum Analysis.
- Any additional resources required. It is not immediately clear why the contractor should have to include this item.
- A method statement showing how the instruction is to be carried out.
- The extension of time required and, presumably to check errors, the new completion date.

- The amount of any loss and/or expense not included in any other estimate.

The contractor and the employer are to take all reasonable steps to agree the estimates. Once agreed, the estimates are binding. That means, for example, that if the contractor has estimated that he will need an extension of time of two weeks, but the effect of the instruction is actually three weeks, the contractor must bear the effect of his bad judgment. That will usually mean that he will have to try and accelerate at his own cost or he must pay liquidated damages for the overrun period.

This agreement is very important and it should be put in writing. If it is a matter of a straightforward agreement of the contractor's estimates by the employer, the simplest method is for the employer to write to the contractor confirming his agreement to the contractor's estimate. If the agreement is to some modified version of the contractor's estimates, it is suggested that the employer should make a counter-offer which the contractor can accept in writing. The parties have ten days in which to reach agreement. They can, of course, agree to extend this period if agreement seems to be near. However, they should take care that agreement is reached as quickly as possible or the contractor's progress may be seriously impeded. Failure to agree in ten days presents the employer with two options:

- He may instruct the contractor to comply with the instruction stating that S6 will not apply; in other words, clauses 12, 25 and 26 will apply in the usual way.
- He may withdraw the instruction. The employer is not liable for any costs incurred by the contractor except costs which are incurred in carrying out design work necessary for the preparation of the estimates. Such design work is to be treated as if it is the result of a change instruction.

If the contractor fails to provide estimates or fails to provide them on time, the consequences for him are rather grim. The instruction is to be valued under clause 12, presumably under 12.4.2, alternative B, because alternative A refers to the price statement which is obviously the sort of thing the contractor is not prepared to produce; however, the point is not certain. Extensions of time are to be made under clause 25 and the amount of loss and/or expense is to be ascertained under clause 26, but the results of the application of clauses 12.4 and 26 will not be received by the contractor until inclusion in the final account and final statement at the end of the contract. Moreover, the contractor is not entitled to any interest or financing charges incurred prior to the issue of the final account and final statement.

10.10 Fluctuations

10.10.1 Choice of fluctuation clause

Clause 35 briefly provides that fluctuations are to be dealt with in accordance with whichever of three clauses is stated in the appendix. It is refreshing to see that if no clause is stated, clause 36 is to apply, thus avoiding any uncertainty.

- Clause 36: Contribution, levy and tax fluctuations used when minimum fluctuations are desired.
- Clause 37: Labour and materials cost and tax fluctuations – used where full fluctuations are intended.
- Clause 38: Use of adjustment formulae – used alternatively where full fluctuations are desired.

These clauses are very similar to the equivalent clauses for use with JCT 98.

Clause 37: Labour and materials cost and tax fluctuations

This is the full fluctuations clause where it is not desired to use the formula. The clause is intricately drafted to achieve its effect. Put as simply as possible, the adjustments are considered in four categories:

- Rates of wages
- Contributions, levies and taxes
- Materials, goods, electricity and fuels
- Landfill tax.

Clause 37.1.1 provides that the prices in the contract sum and the Contract Sum Analysis are based on the rates of wages, other emoluments and expenses payable by the contractor to or in respect of workpeople on the site or off-site, but directly employed and engaged on production for the Works. The rules or decisions of the National Joint Council for the Building Industry or other wage-fixing body current at base date are to apply, together with any appropriate incentive scheme and the terms of the Building and Civil Engineering Annual and Public Holidays Agreements and the like. The prices also take into account the rates of contribution, levy or tax payable by the contractor in his capacity as employer and include such things as the cost of employer's liability insurance, third party insurance and holiday credits.

Clause 37.1.2 stipulates that increases or decreases in the rates of wages or other emoluments and expenses due to alterations in the rules, etc. after base date must be paid to or allowed by the contractor together with consequential increases or decreases in such things as employer's liability insurance, contributions, levies and taxes, etc. Clause 37.1.3 provides that in respect of persons employed on the site, but not defined as workpeople, the contractor must be paid or must allow the same amount as payable in respect of a craftsman under clause 37.1.2. 'Workpeople' is defined in clause 37.7.2 as persons whose rates of wages and other emoluments are governed by the relevant bodies noted in clause 37.1.1. Clause 37.1.5 provides for adjustment in respect of increases or decreases in reimbursement of fares covered by the rules of the appropriate wage-fixing body and if the transport charges in the basic transport charges list are increased or decreased.

Clause 37.2.1 provides that with regard to contributions, levies and taxes, the prices contained in the contract sum and the Contract Sum Analysis are based on the types and rates of contribution, levy and tax payable by the contractor in his capacity as an employer. Again the datum is the types and rates payable at base date. 'Contributions, levies and taxes' is defined very broadly in clause 37.2.8 as all impositions payable by the contractor as employer providing that they affect the cost to the employer of having persons in his employment. In general terms, changes or cessation in payments after base date are allowable as fluctuations.

Clause 37.3.1 provides that the prices in the contract sum and the Contract Sum Analysis are based on market prices of materials, goods, electricity and, if the Employer's Requirements expressly permit, fuels, specified in a list attached to the Contractor's Proposals and current at the base date. Changes in the market prices after the base date are allowable as fluctuations.

Clause 37.3.4 provides that the incidence and rate of landfill tax on waste put on a licensed site are included in the contract sum. If the rate of such tax is increased or decreased after the base date, the amount is to be paid to or allowed by the contractor. There is a stipulation that no payment will be made if the contractor could reasonably have been expected to dispose of the waste material without using a licensed site.

Clause 37.4 provides that the contractor must incorporate provisions having the same effect if he sublets any of the Works. Any increase or decrease in the price payable under the subcontract and which is due to the operation of such incorporated provisions is to be paid to or allowed by the contractor.

Clauses 37.5 to 37.7 are mainly concerned with procedural matters. Importantly, certain notices are to be given by the contractor. Each notice is expressed as being a condition precedent to payment being made to the contractor in respect of the event of which notice is to be given. Clause 37.5.4 deals with adjustments. The contract sum may be adjusted and adjustments under clause 37 must be taken into account when the determination provisions are implemented. The contractor is to provide such evidence and computations as the employer may reasonably require to enable the amounts to be ascertained. That is not to say that the employer is responsible for calculating fluctuations. Indeed, clause 37.5.3 states that the parties may agree on the matter. In practice, the contractor will normally carry out the appropriate calculations and the Employer's Requirements should require the inclusion of all relevant calculations and evidence at the time of submission of applications for payment. Clause 37.5.7 states that if the contractor is in default as regards time, recovery of fluctuations is frozen when he becomes in default. This is subject to two very important provisos:

- There must be no amendment to the printed text of clause 25 (extensions of time).
- The employer must respond to every written notice of delay from the contractor by fixing a new completion date or confirming the old one. This must be done in writing.

These are points which can easily be overlooked when the contract is being set up or during the progress of the work. These fluctuation provisions are expressly not to apply to daywork rates or to changes in rate of VAT.

Clause 36: Contribution, levy and tax fluctuations

This clause is similar to clause 37, but it is limited to duties, taxes and the like and excludes fluctuations in the prices of labour or materials. This is the minimum fluctuation clause under this form of contract.

Clause 38: Use of price adjustment formula

This clause is used where full fluctuations are to be dealt with by the use of formulae. Adjustment is to be made in accordance with this clause and the formula rules for use with clause 38 of the Standard Form of Building Contract With Contractor's Design issued by the JCT and current at the base date. If this system is to be used, it is essential that the Employer's Requirements request the Contract Sum Analysis in the proper form. Helpful guidance is given in Practice Note 23.

10.11 VAT

Clause 14 deals with Value Added Tax and it is similar to clause 15 of JCT 98. Clause 14.1 refers to the supplemental provisions, called the 'VAT Agreement', which are annexed to the contract. The contract sum is exclusive of VAT and the contractor is entitled to recover from the employer any VAT which he has to pay on goods or services. The parties are required to state in the appendix whether or not clause 1A of the VAT Agreement applies. Clause 1A applies only if the contractor is satisfied at the date the contract is executed that his output tax on all supplies to the employer under the contract will be either a positive or a zero rate of tax.

CHAPTER ELEVEN
INSURANCE AND INDEMNITIES

11.1 *Injury to persons and property*

The indemnity and insurance provisions contained in clauses 20, 21 and 22 were extensively revised by amendment 1 issued in November 1986 and again to a limited extent by amendment 10 in July 1996. They are almost identical to the insurance and indemnity clauses in JCT 98. Where the employer employs an agent, it will be the agent's duty either to check the contractor's insurance for the employer, to get an expert to do so or to make sure that the employer seeks advice from his own expert: *Pozzolanic Lytag Ltd* v. *Bryan Hobson Associates* (1998).

Under clauses 20.1 and 20.2 the contractor assumes liability for and indemnifies the employer against any liability arising out of the execution of the Works in respect of the following:

- Personal injury or death of any person except to the extent that the injury or death is due to any act or neglect of the employer or anyone for whom the employer is responsible including directly employed contractors under the provisions of clause 29. The employer's agent will also be included. In the case of personal injury or death, therefore, the contractor is liable unless some or all of the blame can be laid at the employer's door. In that case, the contractor's liability is to be reduced in proportion to the employer's liability. Thus, if the employer is 30% to blame, the contractor will be liable for the remaining 70%.
- Injury or damage to any property to the extent that the injury or damage is due to negligence, breach of statutory duty, omission or default of the contractor or any person for whom he is responsible or who may be on the site in connection with the Works except for the employer and persons for whom he is responsible, local authority or statutory undertaking. In this case, the contractor is not liable unless he can be shown to be at fault and even then, he is liable only to the extent it can be so demonstrated. The contractor's liability does not apply to loss or

damage caused by specified perils to property for which clause 22C.1 insurance has been taken out by the employer. Neither does it apply to the Works or materials on site up to the date of practical completion, partial possession of a particular part or determination.

There is a striking difference between the two indemnities. In the first one, the contractor is liable unless the employer is shown to be liable. In the second case, the contractor is only liable so far as it can be shown that he is in default. The indemnity clauses are important, because they establish the occurrences for which the contractor must assume liability or partial liability. Indemnity clauses are always strictly interpreted by the courts. In particular, it requires very clear words for a person to be indemnified against the consequences of his own negligence: *Walters* v. *Whessoe* (1960). It is not thought that these clauses fall into that category.

Clause 21.1.1 provides that the contractor must take out and maintain insurance against his liabilities under clauses 20.1 and 20.2. This duty is stated to be without prejudice to the contractor's obligation under those clauses. Thus, the taking out of insurance does not affect the contractor's liabilities. If the contractor fails to take out or maintain the cover or if the insurer refuses to meet a claim, the contractor will have to find the money himself. The insurance must be for a sum of money which is not less than the sum stated in appendix 1 for any one occurrence or series of occurrences arising out of one event.

Where the insurance refers to the personal injury or death of one of the contractor's employees arising in the course of employment, it must comply with all relevant legislation.

The employer has the right to require the contractor to provide evidence that the insurances have been taken out and are being maintained. The only stipulation is that the request must be reasonable. In addition, the employer may ask for the policies and premium receipts at any time provided that the request is neither unreasonable nor vexatious. It is not entirely clear what the JCT had in mind when drafting that particular provision, which seems to be repetitive. The employer is given important powers by clause 21.1.3. If the contractor does not take out or maintain the appropriate insurances under clause 21.1.1, the employer may take out the insurance himself deducting the premium cost from any money due to the contractor, after first serving the appropriate notice, of course. Before the employer can take this step, he must establish that the contractor is in default. Although the contract does not specify any time by which the contractor must have effected insurance, in the context of the contract as a whole, the employer and/or the contractor will be in considerable financial danger if the insurances are not in place before the contractor takes possession of the site. A wise employer will ensure that he requests details of insurances from the contractor under clause 21.1.2 before possession takes place.

11.2 *Employer's liability*

Clause 21.2 provides for insurance against damage caused by the carrying out of the Works, when neither contractor nor employer are negligent or in default in any way. This loophole was highlighted in *Gold* v. *Patman & Fotheringham Ltd* (1958) following which the predecessor of this clause was introduced into earlier standard forms. The clause is operative only where it is stated in the Employer's Requirements that the insurance is required. It is not clear why this statement is not included in appendix 1 as is the case with JCT 98, particularly as the amount of indemnity is to be stated there. If the insurance is required, the contractor must take out and maintain the insurance in joint names. A footnote states that the expiry date should not be before the end of the defects liability period. By taking out insurance in joint names, both contractor and employer are the insured. This is important in respect of this type of insurance, because the employer may often be the one to receive a claim following damage to other property. If the contractor simply took out this insurance in his own name, the insurer would not be liable to cover any loss unless the claim was made against the contractor. In the broader context of the later insurance provisions, insuring in joint names ensures that if the loss is due to some negligence of either contractor or employer, the insurer cannot simply pay the insured party and then use his rights of subrogation to recover the payment from the party in default. Subrogation is the right of an insurer to stand in the place of the insured receiving payment in order to pursue legal action in the insured's name against the defaulting party.

The insurance must be taken out against any expense, liability, loss, claim or proceedings which the employer may suffer as a result of damage to any property from certain specified causes:

- Collapse
- Subsidence
- Heave

- Vibration
- Weakening or removal of support
- Lowering of ground water.

There can be few contracts where some degree of insurance of this kind is not required, because of the lack of green field sites. In practice, this kind of insurance covers situations which are common in town and city centre sites. Typically, the contractor will excavate near to adjacent property, or perhaps he will be inserting piles. Although the contractor takes every reasonable precaution and carries out the work with exemplary precision, the adjacent property suffers damage because the walls are inadequately founded or a peculiarity of the underlying strata sets up a vibration which the property cannot resist. In other words, none of the parties concerned were at fault and the danger could not be anticipated - what is commonly called a pure accident. Although no one is at fault, the adjacent owner will certainly have an action against the employer and/or the contractor in the tort of nuisance. The insurance does not cover the following damage:

- If the contractor is liable under clause 20.2 (injury or damage to property real or personal), because this clause is principally for the benefit of the employer, and the contractor indemnifies the employer under clause 20.2 and insures the risk under clause 21.1.
- If it is due to errors in the design. This presumably refers to both the design, if any, provided by the employer through his professional advisors as part of the Employer's Requirements, and to the completion of the design by the contractor either directly or through his subcontractor.
- If it is reasonably clear that damage will result from the building operations. This is a difficult exception, because very many operations in town centre sites almost inevitably cause some damage to adjacent property even if it is only slight cracking. It may depend on just what the insurers are prepared to cover after hearing expert opinion.
- If the employer must insure against just that kind of damage as part of the insurance under clause 22C.1 (insurance of existing structures).
- To the Works and materials on site.
- If the damage results from war, hostilities or the like.
- If the damage is caused by any of the excepted risks, which are listed in clause 1.3 and cover such things as ionising radiations, nuclear risks and sonic booms.

- Caused by or arising from pollution unless occurring as a result of one incident.
- If it results in the employer being liable for breach of contract unless the damage would have been applicable without a contract.

Insurance under this clause must be placed with insurers to be approved by the employer. The contractor is not merely to produce the policy on request; he must deposit the policy with the employer. If the contractor defaults, the employer has power under clause 21.2.3 to take out the appropriate insurance himself and to deduct the cost of premiums from any sums payable to the contractor. It is essential, if this type of insurance is required, that it is taken out before the contractor takes possession of the site.

Clause 21.3 makes clear that the contractor has no liability to indemnify the employer or insure against personal injury, death or injury or damage to property including the Works and site materials caused by any of the excepted risks.

11.3 *Insurance of the Works*

Clause 22 deals with insurance of the Works. The scheme is relatively simple in principle although fairly complex in detail. It is virtually identical to the Works insurance provisions in JCT 98. There is a choice of three clauses:

(1) Clause 22A if new buildings are to be insured by the contractor
(2) Clause 22B if new buildings are to be insured by the employer
(3) Clause 22C when the work is in, or extensions to, existing buildings.

One of these clauses must apply as stated in appendix 1. Care must be taken to choose the correct clause. For example, if clause 22A was stated to apply when the contract called for work to an existing building, the employer may find himself partially or totally uninsured and liable to foot the bill for any damage. Two categories of insurance are involved: all risks and specified perils.

All risks insurance is a very broad category of insurance which covers any physical loss or damage to work executed and site materials with certain exceptions set out in the clause. 'Site materials' as noted in this definition is a term which the contract defines in clause 1.3 as 'all unfixed materials and goods delivered to, placed on or adjacent to the Works and intended for incorporation therein'. Practice Note 22 states that the main additional risks covered by all risks insurance to those covered by the former 'clause 22 perils' are impact, subsidence, theft and vandalism. Clause 22 perils have now been replaced by specified perils.

Specified perils insurance is restricted to the items in the definition in clause 1.3 and includes such things as fire, flood and earthquake, but it excludes the excepted risks.

Clause 22.3 is a very important clause for subcontractors. It places an obligation on the contractor or the employer, whoever has the duty to insure, to ensure that the joint names policies mentioned in clauses 22A.1, 22A.3, 22B.1 or 22C.2 have certain safeguards for any subcontractor referred to in clause 18.2.1. It is to be noted that this clause deals with subletting the work. Where design is sublet under clause 18.2.3, such subcontractors do not appear to benefit from clause 22.3. The latter category will normally include only the professional design team. Where a subcontractor carries out design and executes work, he will benefit.

Either contractor or employer as appropriate must ensure that the policy either recognises each subcontractor as an insured, or it must include a waiver of subrogation by the insurers. The protection applies only in respect of losses by the specified perils to the Works and site materials and it is to last until practical completion or determination. Either the contractor or the employer must ensure these safeguards are included in any policies which are taken out following default of the other party. The value to a subcontractor is that, for example, a subcontractor who causes damage to the Works by fire following his negligence will not be open to any action by the insurers to recover money paid out to the employer or the contractor. The protection does not extend to joint names policies taken out in respect of existing structures and contents under clause 22C.1. In such cases, the insurer will retain subrogation rights and the subcontractors owe a duty of care to the employer so as to be liable in damages if they negligently cause one of the specified perils: *British Telecommunications plc* v. *James Thomson and Sons (Engineers) Ltd* (1998).

11.4 Insurance of the Works: new building

The contractor's obligations are set out in clause 22A.1. He must take out and maintain a joint names policy for all risks insurance.

The value must cover full reinstatement of the works and the amount of any professional fees inserted in appendix 1. Professional fees are the fees required by the professionals involved in the reconstruction work. The employer is entitled to deduct from insurance proceeds the amount incurred for professional fees. A strict reading of the wording suggests that if the percentage is omitted from the appendix, the employer would be obliged to pay the fees himself. The reinstatement value should be carefully considered. If the building is effectively a total loss at a point when 50 per cent of the work has been completed, the cost of demolition of what remains together with reconstruction at inflated prices could result in the contractor having to subsidise the project. The contractor should get very good advice from his broker before taking out insurance to cover this risk. It does not include consequential loss: *Kruger Tissue (Industrial) Ltd* v. *Frank Galliers Ltd and DMC Industrial Roofing & Cladding Services and H & H Construction* (1998). The insurance must be maintained until practical completion of the Works or determination of the contractor's employment even though such determination may be the subject of dispute between the parties. The employer must approve the insurers and he is entitled to have the policy documents and premium receipts. If the contractor defaults, the employer has the right to take out the policy himself and deduct the cost from any sums payable to the contractor.

In practice, of course, the contractor will usually maintain an annual policy which provides cover against all the risks which he may face. So the one policy, possibly by endorsements, will include cover against liability for injury or death to persons, injury or damage to property other than the Works, employer's liability and Works insurance. Clause 22A.3.1 makes provision for this situation and allows the contractor to discharge his obligations under clause 22A provided that the policy is in joint names (a separate endorsement is required for each contract undertaken) and that it provides cover for not less than full reinstatement and professional fees. The contractor must provide evidence that the insurance is being maintained if the employer so requests, but there is no obligation to deposit the policy. The annual renewal date is to be inserted in appendix 1.

The mechanics of the clause are contained in clause 22A.4. If any loss or damage occurs, the contractor must give written notice to the employer as soon as he discovers the loss. A very important provision (clause 22A.4.2) makes clear that the fact that part of the Works has been damaged must be ignored when the amount payable to the contractor is being calculated. He must be paid for the work carried out although it may since have been destroyed. There should be no problems in this respect if payment is being made on a periodic basis (see section 10.2.2); the next payment application will be made and paid as usual. If payment is to be made by stages, there could be difficulties as the payments are not to be made until completion of each stage (clause 30.3.1.1). Thus, the contractor will be paid, but he may have to wait for some time before he can make application for it, particularly if the damage is extensive.

Clause 22A.4.3 has been known to cause difficulties. It places a duty on the contractor to carry out restoration and remedial work and proceed with the Works when the insurers have carried out any inspection they require. It may take the insurers a considerable time to accept the claim. The contractor is not entitled to wait until he knows whether or not the claim will be accepted before he proceeds with the Works. The result is often a heavy financial burden on the contractor. If the damage is very serious, the insurers may employ their own engineers and surveyors to assess the feasibility of repair or total reconstruction. The contract makes no provision for this, but it would be an extremely foolhardy contractor who proceeded with his own ideas of reconstruction in the face of the insurers' own views. It should also be noted that a contractor is not entitled to any extension of time if the cause of the damage lies outside those items listed under specified perils. For example, if the building shell was erected and subsequently collapsed, the contractor would receive no extension of time for the resulting delay no matter who was ultimately at fault. When serious damage occurs, it is in the interests of both parties to obtain first class advice.

The contractor and all his subcontractors who are recognised as insured must authorise the insurers to pay insurance monies to the employer. The contractor is entitled to be paid all the money except for any percentage noted in appendix 1 for professional fees. It is thought that the effect of the wording is that the employer may retain only the amount he has paid out or is legally obliged to pay out in professional fees directly related to the loss or damage, but that there is a ceiling on the amount he may retain. That ceiling is set by the percentage.

The contractor receives the insurance money from the employer in instalments in accordance with clause 30, alternative B (periodic payments), even if the mode of payment stipulated in the contract is actually alternative A (stage payments). The contractor is not entitled to receive more than the insurance money and, if there has been an element of underinsurance or the policy carries an excess or if the insurers repudiate their liability, it is for the contractor to make up the shortfall.

Clause 22B provides for new building insurance to be taken out by the employer. This is not common in practice. The obligation is principally contained in clause 22B.1 and it is similar to the contractor's duties under clause 22A.1. The employer must take out and maintain a joint names policy for all risks to cover the full reinstatement value of the Works together with the percentage to cover professional fees. The employer must maintain the policy until practical completion or determination, whichever is earlier. The employer must produce evidence for the contractor that the policy has been taken out and, on default, the contractor may himself take out a similar policy and he may recover the cost as an addition to the contract sum.

Clause 22B.3 sets out the machinery for dealing with an insurance loss. It closely follows clause 22A.4 and provides for the contractor to give written notice to the employer upon discovering loss or damage. The contractor must proceed with repairs and the execution of the Works after any inspection required by the insurers, and the contractor and his subcontractors who are noted as insured must authorise payment of insurance monies directly to the employer. Here, however, the similarity ends. Where the employer has insured, clause 22B.3.5 stipulates that restoration, replacement and repair must be treated as if they were a change in the Employer's Requirements. There are two important points to note. First, the change does not depend on an instruction from the employer. The fact that there has been loss or damage and the employer has the obligation to insure is sufficient. Second, it follows that if the repairs, etc. are to be treated as a change, they are to be valued and the employer must pay for them. This duty is not affected by any shortfall or excess in the employer's insurance nor is it affected if the insurers decide to repudiate liability. Under 22B, it is the employer who must make good any shortfall.

11.5 *Insurance of the Works: existing building*

If work is to be undertaken in an existing building or in extensions to an existing building, the appropriate clause is 22C. The insurance is to be taken out and maintained by the employer and his obligations are set out in clauses 22C.1 (existing buildings) and 22C.2 (Works in or extensions to existing buildings). Clause 22C.1 refers to a policy in joint names to cover the existing building and contents for specified perils only. The contents are those which are owned by the employer or for which he is responsible. This is presumably intended to cover his goods, goods on the premises with his permission, but not goods which may be on the premises without his permission. Where portions of the new Works are taken into possession by the employer under clause 17, they are to be considered part of the existing building from the relevant date. This is a point which the employer must watch when taking possession of portions of the Works under clause 17. Clause 22C.2 obliges the employer to take out a joint names policy for the new Works in respect of all risks.

Both sets of insurance must be taken out for full reinstatement value, but only in the case of the new Works must the professional fees percentage be added. Both insurances must be maintained until practical completion or determination.

Clause 22C.3 is to be deleted if the employer is a local authority. It gives the contractor broad powers if the employer defaults. The contractor has the usual power to require proof that the insurances are taken out and are being maintained. In addition, in the case of default in respect of clause 22C.1 insurance, he has right of entry into the existing premises to inspect, make a survey and make an inventory of the existing structures and the contents. The only qualification on the contractor's power is that the right of entry and inspection is such as may be required to make the survey and inventory. This provision merits careful consideration by the employer, because a failure to insure by the employer may give rise to distinctly unwelcome, but lawful, entry by the contractor into the employer's property.

The machinery for dealing with loss or damage is set out in clause 22C.4. Note that there is no express machinery for dealing with damage to the existing building and contents. The contractor must give written notice to the employer on discovery. The contractor and his subcontractors noted as insured must authorise the insurers to pay any insurance money directly to the employer. There is provision for either party to determine the contractor's employment within 28 days of the occurrence if it is just and equitable to do so (see section 12.5). If there is no determination or an arbitrator decides that the notice of determination should not be upheld, the procedure is much the same as clause 22B.3. The contractor is obliged to proceed after any inspection required by the insurers, but the work is to be treated as a change for which the employer must pay. Shortfalls in insurance under this clause are again the responsibility of the employer. Under none of the three Works insurance clause options is the contractor penalised in respect of work already carried out and damaged by the insurance risk. It has been held that fire caused by the contractor's negligence is not covered by this insurance, but must be covered by the contractor's own insurance: *London Borough of Barking and Dagenham* v. *Stamford Asphalt Company* (1997). The case dealt with the JCT Minor Works Contract (MW 80), but there seems no reason why the principle should not be applied to WCD 98. However, it does cut across generally accepted principles that the insurance is to deal with damage however it may be caused. The case seems to call into question one of the purposes of insurance in joint names.

11.6 Employer's loss of liquidated damages

Clause 25 allows the employer to make an extension of time if the contractor is delayed to the extent that the date for completion is exceeded by loss or damage caused by one or more of the specified perils (clause 25.4.3). The employer is then faced with the prospect of receiving his building after the date for completion, but he will not receive liquidated damages for the delay because the contractor has received an extension of time. Clause 22D is intended to provide the employer with some relief in these circumstances. More cynically, it is simply more insurance which the employer can take out on payment of the appropriate premium. The insurance pays out if extensions are given on grounds of delay due to specified perils.

For this clause to apply, it must be so stated both in the Employer's Requirements and in appendix 1. As soon as he reasonably can, after the contract has been entered into, the employer must notify the contractor that no insurance will be required. Alternatively, he must instruct the contractor to obtain a quotation and for that purpose provide whatever information the contractor reasonably requires. The contractor must act as soon as reasonably practicable to send the quotation to the employer, who must instruct acceptance or otherwise. If the employer wishes to accept, the contractor must deposit the insurance policy, together with premium receipts, with the employer.

The insurance is to be on an agreed value basis, the value to be the amount of liquidated damages at the rate stated in appendix 1. The purpose of 'agreed value' is to avoid arguments with the insurers when called on to pay out, because they would normally expect to pay only the amount of actual loss. A footnote to the clause points out that insurers will normally reserve the right to be satisfied that the amount of liquidated damages does not exceed a genuine pre-estimate of the damages which the employer considers he will suffer. The period of liquidated damages must also be stated in appendix 1. Thus, if the stated period is four weeks and the liquidated damages are set at £100 per week, an extension of time of two weeks for relevant event 25.4.3 will enable the employer to claim £200 from the insurers. If the extension is for five weeks, he can only claim £400, because four weeks is the limit of insurance.

Clause 22D.4 permits the employer to take out the insurance himself if the contractor defaults.

11.7 *The joint fire code*

The joint fire code is dealt with under clause 22FC. Clause 1.3 defines the joint fire code as:

> 'the Joint Code of Practice on the Protection from Fire of Construction Sites and Buildings Undergoing Renovation which is published by the Building Employers Confederation (now Construction Confederation), the Loss Prevention Council and the National Contractors' Group with the support of the Association of British Insurers, the Chief and Assistant Chief Fire Officers Association and the London Fire Brigade which is current at the Base Date.'

The code makes clear that non-compliance could result in insurance ceasing to be available. If the code is to apply, the appendix should be completed appropriately. If the insurer categories the Works as a 'Large Project', special considerations apply and the appendix must record that also.

Clause 22FC.2 requires both employer and contractor and anyone employed by them and anyone on the Works including local authorities or statutory undertakers to comply with the code. Clause 22FC.3 makes clear that if there is a breach of the code and the insurer gives notice requiring remedial measures, the contractor must ensure the measures are carried out by the date specified in the notice and in accordance with the architect's instructions if any. If the contractor does not begin the remedial measures in seven days from receipt of the notice, or if he fails to proceed regularly and diligently, the employer may employ and pay others to do the work. In principle, the employer is entitled to recover the cost by deduction or as a debt in the usual way. Clause 22FC.4 provides that the employer and the contractor indemnify each other against the consequences of a breach of the code. Therefore, if, although the contractor did not carry out the remedial measures, the cause of the breach lay at the door of the employer, he could not recover the cost from the contractor. If the code is amended after the base date and the contractor is put to additional cost in complying, such cost must be added to the contract sum.

CHAPTER TWELVE
DETERMINATION

12.1 *Common law position*

Under the general law, a contract can be brought to an end in four main ways:

- By performance
- By agreement
- By frustration
- By breach and its acceptance.

12.1.1 Performance

This is the ideal way of bringing a contract to an end when both parties have carried out their obligations under the contract and nothing further remains to be done. At that point, the purpose for which they entered into the contract has been accomplished and the contractual relationship ceases.

12.1.2 Agreement

If the parties to a contract so wish, they may agree to bring the contract to an end. What they are actually doing is entering into another contract whose sole purpose is to end the first contract. In most cases, when a contract is ended by mutual agreement it is because the parties gain something from so doing, thus satisfying the requirement for consideration as an essential element of the contract. However, it is prudent for the parties to execute the second contract as a deed, thus avoiding any question of consideration arising.

12.1.3 Frustration

A useful definition of frustration was given by Lord Radcliffe in *Davis Contractors Ltd* v. *Fareham Urban District Council* (1956):

> '(It) occurs wherever the law recognises that without default of either party a contractual obligation has become incapable of being performed because the circumstances in which performance is called for would render it a thing radically different from that which was undertaken by the contract.'

A straightforward example of a contract being frustrated is if a painting contractor entered into a contract to repaint the external woodwork of a house and before he could commence work, the house was destroyed by fire which was neither the responsibility of the building owner nor the painter. There are other cases where it will be a question of degree whether the contract is frustrated. The fact that a contractor experiences greater difficulty in carrying out the contract or that it costs him far more than he could reasonably have expected is not sufficient ground for frustration. Neither will a contract be frustrated by the occurrence of some event which the contract itself contemplated and for which it made provision: *Wates Ltd* v. *Greater London Council* (1983). In practice, it is very rare for a contract to be frustrated.

12.1.4 Breach

A breach of contract which is capable of bringing the contract to an end is termed a repudiatory breach. A breach of this nature must strike at the very root of the contract: *Photo Production Ltd* v. *Securicor Ltd* (1980). The offending party must clearly demonstrate that he does not intend to accept his obligations under the contract. Such an instance under a building contract could take place where the employer, growing weary of what he perceived as continuing and irredeemable faults on the part of the contractor, prevents all further access onto the site and engages another contractor to complete the work. That would be a very clear repudiation by the employer. Many acts of repudiation are less obvious.

There is no common law right for any party to treat a contract as repudiated simply because the other party is in breach of his obligation to pay. Consistent failures to pay, however, such that a party has lost all confidence of ever being paid may be repudiation in certain circumstances: *D.R. Bradley (Cable Jointing) Ltd* v. *Jefco Mechanical Services* (1988).

A repudiatory breach by one party does not automatically end the contract. The innocent party has the right to affirm the contract and claim damages arising from the breach, or he may accept the breach, bringing the contract to an end, and claim damages.

12.2 *Determination generally*

The contractual provisions for determination do not provide for one or other of the parties to bring the contract to an end, because to do so would mean that all the clauses in the contract (with the exception of the arbitration clause: *Heyman* v. *Darwins Ltd* (1942)) would fall. The contract expressly provides for determination of the contractor's employment under the contract. This puts beyond all doubt that the clauses dealing with consequences of determination continue to apply. Although there is no doubt that 'determination' has the meaning of 'cessation' or 'conclusion', it also means 'fixing' or 'definition' or 'exact ascertainment'. It is a term used in all JCT contracts, but it might have been in the interests of wider understanding to have referred to 'termination'.

The determination clauses in WCD 98 are very similar, but not identical, to the equivalent clauses of JCT 98. The similarity induces a misplaced familiarity which, in turn, can lead to difficulties. They were substantially varied by JCT Amendment 7.

It is essential to remember that the grounds for determination under this contract would not all amount to repudiatory breaches at common law. It is useful to have a specific contractual machinery for determination because to rely on common law repudiation can be very uncertain. However, contractual determination must not be thought to end all problems in that respect, because deciding whether the precise grounds have been satisfied can bring its own problems of interpretation of the clauses and of the facts.

In certain instances, where the facts give the innocent party the choice, it may pay him to accept the breach as repudiatory at common law rather than proceed to operate the determination mechanism under the contract. This is because the acceptance of repudiation entitles the party to damages, whereas to determine under the contract simply entitles the party to whatever remedies the contract stipulates. This may not always be sufficient: *Thomas Feather & Co (Bradford) Ltd* v. *Keighley Corporation* (1953). In rare cases, a party may be able to rely on the contractual determination provisions and acceptance of repudiation in the alternative: *Architectural Installation Services Ltd* v. *James Gibbons Windows Ltd* (1989).

The principal determination clauses under this contract are: 27, 28 and 28A. Determination may also occur under clause 22C.4.3.1.

12.3 *Determination by the employer*

12.3.1 Grounds

Clause 27 sets out the terms on which the employer may determine the contractor's employment. It is expressed as being without prejudice to any other rights and remedies which the employer may have. That means that the employer is not confined to using the determination clause; he may also treat the contract as ended by accepting the contractor's repudiatory breach if such a course suits the employer's interests better.

There are seven grounds for determination, five of which are based on the contractor's default. These five are as follows.

Suspension of design or construction

Because of the particular nature of this contract, the equivalent ground under JCT 98 has been expanded to include design. In practice, it will be difficult to identify suspension of design work. Construction can be observed progressing (or not progressing) on the site. Design normally takes place in an office somewhere, quite possibly at some distance from the contractor's own office. The employer would usually only know that design work was suspended if he was specifically informed. In order to qualify under this ground, the suspension must be complete or substantial. It must also be without reasonable cause. It could be said that it would be reasonable for the contractor to suspend work if he was waiting for some consent or approval which the employer must give. The contractor will be entitled to suspend for failure to pay under the Housing Grants, Construction and Regeneration Act 1996. This power is exercised under the contract clause 30.3.8, which expressly excludes it from treatment as a suspension under this ground. Whether the contractor's suspension was reasonable would depend on the facts in each case. It would certainly be dangerous for the employer to rely on the contractor's suspension as ground for determination if the contractor could put up a convincing argument, even if suspension for that reason was not strictly in accordance with the contract. To be reasonable in this context, it may not be necessary to be empowered by the contract.

Failure to proceed regularly and diligently

The meaning of 'regularly and diligently' has been considered in section 7.2. This ground is a breach of the contractor's obligation to comply with clause 23.1.1. Guidelines were laid down in *West Faulkner Associates* v. *London Borough of Newham* (1995) which help to show whether the contractor is failing in this regard. Suspension under clause 30.3.8 is not a failure to proceed regularly and diligently for the purposes of this ground.

Failure to remove defective work

A number of criteria must be satisfied before this ground can be invoked:

- The employer must have given a written notice to the contractor; *and*
- The notice must require the removal of defective work, materials or goods; *and*
- The contractor must have refused or neglected to comply; *and*
- As a result the Works must be substantially affected.

This ground relates to breach of an instruction given by the employer under clause 8.4.1. At first sight, determination appears to be a draconian measure in response, for example, to the contractor's failure to remove defective door handles. It is doubtful whether such a failure would qualify as substantially affecting the Works. Clearly what is indicated is a failure on the part of the contractor which is deliberate ('refuses or neglects') and which seriously affects the Works. Such an instance might occur where the contractor deliberately neglects to remove some defective work which is about to be covered up and possibly may be required to give support to further work and the subsequent rectification will cause significant delay and expense. Even if the contractor appeared reluctant, the employer could use his powers under clause 4.1.2 to engage others to carry out the rectification without determining the contractor's employment.

Failure to comply with the assignment and subcontracting clauses

The contractor's obligations under clause 18 are to obtain the employer's consent before assignment of the contract or subletting. Assignment without consent would be ineffective: *St Martins Property Corporation Ltd and St Martins Property Investments Ltd* v. *Sir Robert McAlpine & Sons Ltd* (1992)). In some circumstances, failure to obtain consent to subletting will be very serious. This is especially the case under this contract where a contractor attempts to sublet the design of the Works without consent. It is suggested that determination will be the last resort in any event.

Failure to comply with the CDM Regulations

This is wider in scope than the relevant parts of clauses 25 and 26 and it effectively gives a determination remedy for failure under the relevant contract clauses to comply with the Regulations. These are the clauses grouped under clause 6A.

12.3.2 Procedure

An employer who intends to determine the contractor's employment must follow the procedure precisely. If the contractor defaults in one of the five ways noted above, the employer must give him a written notice which must specify the default. It is thought that, strictly, the default is one of the five noted above and not the precise circumstances of the default which is more correctly the evidence supporting the allegation of default. For example, the employer need only say that the default consists in failure to proceed regularly and diligently. He need not say that the contractor had only three men on site for a month and that less than a thousand pounds worth of work was carried out in that time, etc. It may not be in the employer's interests to give too much detail at this stage. It is necessary that the contractor is in no doubt about the default alleged and the employer must give sufficient detail to identify the incident if there is any danger of confusion: *Wiltshier Construction (South) Ltd* v. *Parkers Developments Ltd* (1997).

All notices are to be given by actual delivery, special or recorded delivery. The employer should take care to follow the notice procedure precisely.

If the contractor continues the default for 14 days after he receives such notice, the employer may within a further 10 days determine his employment by serving a notice of determination. The employer is not to act unreasonably or vexatiously. 'Vexatiously' suggests an ulterior motive to oppress or annoy. Whether he is unreasonable must be decided by an objective judgment: *John Jarvis Ltd* v. *Rockdale Housing Association Ltd* (1986). This proviso is particularly important when the employer's broader powers are considered.

If the contractor stops the default within the stipulated 14 day period or the employer fails to issue a notice of determination, but subsequently the contractor commits the same default again, perhaps months later, the employer may determine his employment straight away or within a reasonable time of the repetition without the need for a further notice. We would caution that the employer must take great care if attempting to put this provision into effect. The default must be precisely the same default which prompted the employer to serve the original default notice. If that criterion can be met, the employer would be sensible to serve a warning notice to the contractor before issuing the determination notice. Although this is not an express contractual requirement, indeed the contract appears to state the contrary, the serving of a further notice of perhaps only three days duration would serve to counter any possibility that the employer could be said to be acting unreasonably.

12.3.3 Further grounds

There are two further grounds for determination under the terms of clause 27 - insolvency and corruption.

Insolvency

Clause 27.3 allows for determination in any of the following circumstances:

- The contractor makes a composition or an arrangement with his creditors
- The contractor makes a proposal for a voluntary arrangement for composition of debts or scheme of arrangement to be approved in accordance with the Companies Act 1985 or the Insolvency Act 1986
- The contractor under the Insolvency Act 1986 has an administrator or an administrative receiver appointed
- The contractor becomes bankrupt
- The contractor has a winding up order made
- A resolution for voluntary winding up is passed (except for the purposes of amalgamation or reconstruction)
- A provisional liquidator is appointed.

What this all amounts to, in simple terms, is that the contractor becomes insolvent. Clause 27.3.3 provides that no notice of determination is required from the employer in the case of the last four grounds and determination is automatic. The employment may be reinstated if the contractor and the employer agree. Legal advice is indicated.

So far as the first three grounds are concerned the employer may determine the contractor's employment by notice at any time and it will be effective on the date it is received. This is subject to clause 27.5.2.1. The employer may decide not to give notice of determination in these cases and exercise other options. Importantly, clause 27.5.1 states that from the date on which the employer could have given a determination notice, he is not bound to make any further payment to the contractor who, in turn, is not bound to continue to carry out and complete the design and construction of the Works.

The employer may take what clause 27.5.4 refers to as 'reasonable measures' to make certain that the Works, materials and the site itself are protected. This may entail putting in place security measures, erecting fences, installing new locks and the like. The contractor must not hinder the employer in carrying out these measures. The employer may deduct the reasonable cost from money due to the contractor (provided the appropriate notice is served) or he may recover it as a debt. Clause 27.5.2 provides that the employer and the contractor may enter into a clause 27.5.2.1 agreement which may be to continue the carrying out of the Works, the novation of the contract or the conditional novation of the contract. From the date of such agreement, the two parties are subject to its terms so far as payment, timing and any other matter regarding the Works are concerned. Obviously, the cessation of payment and performance noted above will make way for new terms.

The only alternative to a clause 27.5.2.1 agreement is for the employer to determine the contractor's employment. Whether it is better to do one or the other in any given set of circumstances will be for the employer to take advice from the employer's agent and from legal advisors. During the period after the employer could have determined, but did not, and the decision to enter into a clause 27.5.2.1 agreement or to determine, clause 27.5.3 gives the employer an interim solution while he is seeking advice and making up his mind. The employer may enter into an interim arrangement for the work to be carried out. The clause usefully provides that during the currency of such arrangement, the employer is not entitled to exercise any right of set-off against any payment due to the contractor under the arrangement except any deduction arising from the taking of reasonable measures under clause 27.5.4.

Corruption

Clause 27.4 provides for determination if the contractor is guilty of corrupt practices. Although it is not expressly stated, the employer should determine by the issue of a written notification. There is no requirement that the employer should serve a preliminary notice, because it matters not that the contractor has stopped the corruption. The employer has the power to determine, but he is not obliged to do so. Although perhaps academic, it is curious that the employer's power is stated to be to determine the contractor's employment 'under this or any other contract'. Presumably this phrase refers to 'any other contract' between the same two parties. Even with that proviso, it is difficult to see how the parties can bind themselves in law under this contract regarding the substance of other contracts. Indeed, we think that they could not do so. It would require a similar clause in other contracts to entitle the employer to determine the contractor's employment under those contracts. Although the employer cannot be given power under this contract to determine the contractor's employment under other contracts, it would be perfectly feasible to insert clauses in this and other contracts to permit the employer to determine the contractor's employment under any contract if he committed some corrupt act in connection with another contract.

The list of qualifying acts is comprehensive:

- If the contractor has offered, given or agreed to give a gift of any kind as reward for doing or not doing any action relating to this contract or any other contract with the employer.
- If the contractor has shown or not shown favour or disfavour in relation to this or any other contract with the employer.
- If such acts have been done by the contractor's employee or agent, whether or not the contractor is aware.
- If the contractor, his employee or agent has committed any offence under the Prevention of Corruption Acts 1889–1916 in relation to this or any other contract.
- If the employer is a local authority and the contractor has given any fee or reward and receiving it is an offence under subsection (2) of section 117 of the Local Government Act 1972.

Corruption is a criminal offence and there are severe penalties. The employer is entitled to rescind the contract at common law and to recover any secret commissions. The clause is very strict and it is to be noted that the contractor cannot plead under the contract that he did not know what was being done. This clause makes clear that if it was done by his men or his agents, he is liable.

12.3.4 Consequences

The consequences of determination are set out in clauses 27.6 to 27.7. They follow determination under any of the clauses 27.2, 27.3 or 27.4. Clause 27.6.1 is a difficult clause. It provides that the contractor must give the employer certain drawings, details, etc. The danger is that the clause will receive little more than a glance in normal circumstances and the employer will assume that, in the event of determination, it entitles him to receive all the information necessary for him to engage some other contractor to finish the job. Careful reading of the clause leads to a different conclusion. The contractor is to provide the drawings, etc. 'for the purposes referred to in clause 5.5'. Clause 5.5 refers to what are commonly termed 'as-built' drawings. They are to concern the maintenance and operation of the Works. If the Works are only half completed at the date of determination, the drawings supplied by the contractor are likely to show very little and certainly nothing of the proposed Works not yet carried out. The reference later in this clause to the drawings which the contractor has prepared and information related to the Works completed before determination reinforces that position. Thus, at determination, the employer is simply entitled to be given and to retain full details of the work which has been carried out. It is not at all certain that the employer would be entitled to retain drawings already provided which show proposals for the remainder of the project.

A contractor could not prevent the employer from using his design by claiming copyright in it, because the employer must have an implied licence (at least) to carry on and complete the Works to the same design in these circumstances. It would be intolerable if a contractor in default could revoke the licence and take advantage of his own default: *Alghussein Establishment* v. *Eton College* (1988). Clause 27.6.2 removes any doubt, because it allows the employer to engage others to complete the design and the construction of the Works and make good defects. They may enter the site and use all the temporary buildings and plant and all materials (subject to any retention of title clause) and purchase any new materials for that purpose. The contractor is not entitled to remove his temporary buildings, etc. from the site until the employer has expressly and in writing required him to do so. The contractor has a reasonable time in which to remove his buildings and plant, etc. If he does not comply with the employer's request, the employer may remove them himself without any liability for damage, sell them and hold the proceeds less costs for the contractor.

Clause 27.6.3 relates to the position of subcontractors and suppliers. It is in two parts. Clause 27.6.3.1 is stated not to apply if the reason for determination was insolvency other than if the contractor was a company and made a proposal for voluntary arrangement for composition of debts or a scheme of arrangement under the Companies Act 1985 or the Insolvency Act 1986. If the employer so requires within 14 days of the date of determination, the contractor must assign the benefit of any agreement for the supply of materials or the carrying out of any work to the employer on terms that the subcontractor or the supplier can object to any further assignment. Indeed, although the clause does not so state, a subcontractor or a supplier could object to assignment of the benefit of their contracts to the employer. The assignment is said to be without payment. That is, without payment to the contractor. Clause 27.6.3.2 does not apply if the contractor has a trustee in bankruptcy, a provisional liquidator, an insolvency petition against it or voluntary winding up resolution. The employer is expressly empowered to pay any subcontractor or supplier for work or materials done or delivered before determination. The only stipulation is that the price has not already been paid by the contractor. No doubt the subcontractors and suppliers will normally be pleased at the prospect of assignment if the employer agrees to pay them what they are owed. The employer may deduct the amount of such payments from the contractor or they may be recovered from him as a debt.

Clauses 27.6.5 to 27.6.7 deal with payment. No further payment or release of the retention is to be made. That prohibition is not to prevent the contractor enforcing his right to be paid any money to which he was due and which the employer has unreasonably not paid and which was payable more than 28 days before the employer could have issued a default notice or, if no default notice, the date of determination.

After making good of defects by the contractor employed to complete the Works, an account must be made by the employer. The account must show expenses properly incurred by the employer together with loss or damage caused as a result of the determination, the amount already paid to the contractor and the amount which would have been payable if the Works had been completed in accordance with the contract. The result may be a debt owing to the contractor or to the employer.

If the employer decides not to continue with the Works, clause 27.7 provides that he must send written notification to the contractor within six months from the determination. This clause responds to the situation which arose in *Tern Construction Group (in administrative receivership)* v. *RBS Garages* (1993), where the judge had to imply a term. The employer must send the contractor a statement of account within a reasonable time of the notice. The statement must set out the total value of work properly executed, any other amounts due to the contractor and the amount of expense properly incurred by the employer together with any loss or damage. After taking into account amounts previously paid, the result may be a debt owing to the employer or to the contractor. If the six months have expired and the employer has neither started to employ others to complete the design and construction of the Works nor sent written notice to the contractor that the Works are not to be completed, the contract gives the contractor certain rights. Clause 27.7.2 stipulates that he may serve written notice on the employer requiring the employer to state whether or not he is to complete the Works. If not, he may require a statement of account to be prepared by the employer. Unfortunately, the contract is silent about the next step if the employer simply does not respond. The contractor would no doubt follow the appropriate dispute resolution procedure.

There is no express provision requiring the contractor to surrender possession of the site on determination, but it is considered that the courts would now grant an injunction if the contractor attempted to stand fast. This view is supported by two useful decisions of Commonwealth courts: *Kong Wah Housing* v. *Desplan Construction* (1991) and *Chermar Productions* v. *Prestest* (1991). These decisions are in sharp contrast to the only English case on the point: *London Borough of Hounslow* v. *Twickenham Garden Developments Ltd* (1970).

12.4 Determination by the contractor

12.4.1 Grounds

Clause 28.2 sets out the grounds on which the contractor may determine his employment under the contract. They are expressed as being without prejudice to any other rights and remedies which the contractor may possess. He is thus placed on the same footing as the employer.

The grounds for determination are divided into two categories.

There are three grounds in the first category, which is to be found in clause 28.2.1.

Employer's failure to pay

The employer must have failed to pay an interim amount properly due. In order for the amount to be properly due it must be contained in an application for payment made under clause 30.3.1 and it must comply with the provisions of clauses 30.1 to 30.4 (see Chapter 10). It should be noted that it is not sufficient for the contractor to make application for some approximated sum and then to determine if it is not paid in full. Unlike a traditional contract, the contractor is at the helm so far as calculating payment is concerned. In these circumstances, if determination is to be effective, the contractor must scrupulously carry out his duties. The valuation must be carefully calculated and the appropriate deductions made. If the employer makes any deduction under clause 30.3.4, the contractor must seriously consider whether such deduction is justified before launching into the determination procedure. This is probably the most common ground for determination and it is certainly valuable, but if the contractor is in error in determining, he could be held to be in repudiatory breach of contract by his actions although much will depend on the extent to which the contractor has honestly relied on the contract provision, even though he may do so mistakenly: *Woodar Investment Development Ltd* v. *Wimpey Construction UK Ltd* (1980). Where the contractor desires to determine on this ground, the employer must have failed to pay within 14 days of the due date. Reference to clause 30.3.3 suggests that the due date is the date of issue of the contractor's application for payment. The date of issue of an architect's certificate has been held to be the date of posting: *Cambs Construction* v. *Nottingham Consultants* (1996).

Employer fails to comply with the assignment clause

Presumably non-compliance would be an attempt to assign without permission which we have already seen would be ineffective.

Employer fails to comply with the CDM Regulations

The comments are the same as under employer determination.

The second category, which is to be found in clause 28.2.2, contains four grounds for determination provided a pre-condition has been satisfied. The pre-condition is that the carrying out of the whole or nearly the whole of the Works has been suspended for the period named in appendix 1 by at least one of the grounds. It is suggested in a footnote to the appendix that the period should be one month. The contract refers to them as 'specified suspension events'. The suspension must be continuous and due to one of the following.

Late instructions

This ground refers to the employer's failure to supply the contractor with necessary instructions, decisions, information or consents at the right time. There are two provisos. First, the employer must be obliged to provide the instruction, etc. under the terms of the contract. Second, the contractor must have specifically applied in writing on a date which is neither too near nor too far away from the date on which it is required. The comments on the contractor's duty to specifically apply under the extension of time provisions apply equally to this clause (see section 8.3). For some reason which is not entirely clear, except perhaps to the JCT, clause 2.4.2 decisions are singled out for mention as being included. This is a decision required from the employer after the contractor has suggested an amendment to overcome a discrepancy in his Proposals. A decision under clause 2.4.1 (discrepancy within the Employer's Requirements) would appear to be equally deserving of mention.

Employer's instructions

Only instructions issued under clauses 2.3.1 (to correct a divergence between the Employer's Requirements and the definition of the site boundary by the employer under clause 7), clause 12.2 (instructions requiring a change in the Requirements) and 23.2 (postponement of design or construction work) qualify under this ground. It is made clear that if the instructions result from negligence or default of the contractor, his servants or agents or any person employed in connection with the Works other than the employer or persons employed or engaged by him, this ground will not bite.

Employer's men

This ground deals with delay in carrying out of work by the employer's directly employed persons or supply of materials which the employer has agreed to supply or the failure to execute the work or to supply the materials. It is difficult to understand why an employer should wish to carry out any work by directly employed labour if he is travelling down the design and build route. By putting his finger into the pie, he is potentially forsaking many of the advantages which this system of procurement offers.

Failure to give access

This ground will apply only if the following criteria are satisfied:

- The access must be set out in the Employer's Requirements and the contractor must have given whatever notice was specified; alternatively, the access must have been agreed between the employer and the contractor.
- The access must be through or over land, buildings, way or passage adjoining or connected with the site.
- The land must be in the possession and in the control of the employer.
- The employer must have failed to give ingress or egress at the appropriate time (e.g. at the date of possession).

(See also section 8.3 for comments on the similar provision as a relevant event for extension of time.)

There are two important points the contractor must watch before determining:

- He must be certain that the suspension has continued for the prescribed period; *and*
- The suspension must be continuous, i.e. there must be no breaks during which work resumed for a period, no matter how brief.

12.4.2 Procedure

If the contractor intends to determine, he must serve notice on the employer specifying the default or specified suspension event and requiring an end within 14 days. The contractor must take care not to give notice too early.

There are certain presumptions which may be made with regard to posting, but such presumptions can be defeated if a party proves that the presumption does not accord with the facts. The contractor can adopt two strategies to overcome the problem. He can send the application by recorded delivery and request confirmation of the date of delivery from the post office. Alternatively, he can deliver the application by hand and obtain a receipt clearly showing the name of the person receiving and the date. The latter option seems more attractive if the employer is not too far away from the contractor's office, because it is certain and it reduces the time period by at least the one day which (theoretically) is required for first class post to reach its destination in the UK.

If the employer does not end the default by the end of the last day of the 14 day period, the contractor may serve a notice on him by actual delivery or special or recorded delivery. The notice should state that notice of determination under clause 28 will be served if the default or suspension is not ended within 14 days from receipt. The notice will expire on the fifteenth day, after which the contractor may serve notice of determination by actual delivery or special or recorded delivery. The contractor has ten days in which to act. If he fails to act within the ten days or if the default or suspension is ended within the 14 day period, the contractor is entitled under clause 28.2.4 to serve notice of determination if at any time the default is repeated or the suspension event is repeated (even for a short period). No warning notice is prescribed in this instance and the contractor may serve the determination notice 'upon or within a reasonable time of such repetition', but it is wise for the contractor to issue a notice in any event. The repetition must be precisely the same as the default or suspension which led to the original default notice.

12.4.3 Further ground - insolvency

Clause 28.3.1 provides for determination for the same insolvency reason as clause 27.3.1. In the event that the employer makes an arrangement or composition with creditors or a company voluntary arrangement, he must immediately inform the contractor in writing when the contractor may determine his employment. The determination takes effect from the date of receipt of the determination notice by the employer. It is important to note that the contractor's obligation to carry out and complete the design and construction of the Works is suspended as soon as any of the insolvency events in clause 28.3.1 occur.

A general proviso at the end of clause 28.1 prohibits the contractor from giving the determination notice unreasonably or vexatiously (see our comments in section 12.3.2).

12.4.3 Consequences

Following determination under clause 28.2 or 28.3, the rights and duties of the parties are set out in clause 28.4. In essence, the situation is straightforward as follows:

- The contractor must remove all his temporary buildings, plant, tools, equipment and materials from site. He must act as quickly as is reasonable in the circumstances and he must take whatever precautions are necessary to prevent death, injury or damage of the types against which he was liable to indemnify the employer under clause 20 before the date of determination (see Chapter 11). The contractor must give all subcontractors facilities to remove their equipment, etc. but neither contractor nor subcontractors may remove goods or materials which have been properly ordered for the Works and for which the contractor has paid or is legally liable to pay. Although the contractor has determined, this provision makes clear that he is not entitled to leave his property on the site until he feels like moving it or until he has another site to receive it. He must act reasonably quickly. Certainly, he will not usually be able to serve notice and move everything away the following day, but after the notice, although not before, he must stop work and commence the process of vacating the site. Other than set out under this clause, the contract clauses requiring release of retention do not apply.
- Clause 28.4.1 refers to the contractor's duty to provide the employer with drawings, etc. for the purposes referred to in clause 5.5. The clause is identical to clause 27.6.1 dealing with the consequences of the employer's determination and our comments on that clause are also applicable here (see section 12.3.4).
- Clause 28.4.3 provides that, within 28 days of the date of determination, the contractor shall be paid the retention subject to the employer's right of deduction providing such right existed before the date of determination. The notice provisions for withholding are particularly important in this regard to establish such right.
- Clause 28.4.4 stipulates that the contractor shall be paid certain amounts after taking account of what he has already been paid under the contract. The clause stipulates that the contractor is to be paid within 28 days of submission by the contractor of his account. The contractor must prepare the account with 'reasonable dispatch'. Clearly he will waste no time. The payments are to consist of the value of all construction work which is properly

executed and all design work carried out (the value to be ascertained under clause 12.4), amounts due for direct loss and/or expense under clauses 26 and 34.3 including amounts ascertained after the date of determination, amounts in respect of materials for which the contractor has paid or is legally liable to pay, the contractor's reasonable costs incurred in removal from site, and direct loss and/or damage which the contractor has incurred as a result of the determination. In an appropriate case, such loss and/or expense could include the whole of the profit which the contractor would have made had he been allowed to complete the contract: *Wraight Ltd* v. *P.H. & T. Holdings Ltd* (1968). This would be subject to the contractor demonstrating that, on the balance of probabilities, he would have made a profit. Claims for loss of profit are not sustainable in the abstract, particularly when all the evidence points to the contractor having made a loss and continuing to do so: *McAlpine Humberoak* v. *McDermott International Inc* (1992).

12.5 *Determination by either party*

12.5.1 Grounds

There are two clauses which allow either party to determine the contractor's employment. They are 28A and 22C.4.3. These clauses deal with situations where neither of the parties is at fault.

Clause 28A

This clause entitles either party to determine the contractor's employment if virtually the whole of the Works is suspended due to any of seven causes. The period of suspension must be continuous for whatever period the parties stipulate in appendix 1. The contract suggests that three months should be stated for the first four causes and that one month is appropriate for the remainder. The causes are:

- *Force majeure*: (see section 8.3).
- Loss or damage due to specified perils: note that neither party is entitled to determine under the broader 'all risks' category. The contractor may not determine under this head if the loss or damage was due to his own negligence or default or to that of those for whom he is responsible or anyone engaged on the

Works other than the employer, those for whom he is responsible, the local authority or statutory undertakers.

- Civil commotion: this is defined as more serious than a riot, but not as serious as a civil war: *Levy* v. *Assicurazioni Generali* (1940).
- Delay in receipt of development control permission. This ground is not to be found in any other JCT contract. Where necessary approvals such as planning permission have not been obtained by the employer before the contract is executed, there is a real risk of the project suffering a long delay before it can commence on site. The contractor must have taken all practicable steps to avoid or reduce the delay, but that probably means little more than that he should have made his application competently and in good time (see also the remarks in section 6.3).
- Employer's instructions issued under clauses 2.3.1, 12.2 or 23.2 as a result of the negligence or default of a local authority or statutory undertaking carrying out statutory obligations. These instructions deal with divergences between the Employer's Requirements and the definition of the site boundary, change instructions and postponement.
- Hostilities involving the UK.
- Terrorist activity.

As soon as the stipulated period has expired, all that is required is for the party intending to determine to serve notice on the other by actual delivery or special or recorded delivery that if the suspension is not terminated within seven days of the date of receipt of the notice, the contractor's employment will determine immediately thereafter.

Clause 22C.4.3

If the employer has insured under clause 22C and there has been loss or damage to the Works, either party may determine the contractor's employment within 28 days after the occurrence provided that it is 'just and equitable' to do so. What is meant by 'just and equitable' is probably to be decided on the facts. It is significant that the power of determination is only given where the Works entail existing structures. It is reasonable to suppose that the clause envisages the situation where the contract is for an extension to an existing building and the Works and the existing building are virtually destroyed. There would be nothing to extend and the employer may wish to consider some other option. It is thought that in most cases where this clause would apply, the contract might well be frustrated. Notice must be given by actual delivery or special or recorded delivery. Either party, on receipt of the notice, has seven days in which to invoke the relevant dispute resolution procedures in order to decide whether the determination is just and equitable.

12.5.2 Consequences

The consequences of determination under clause 28A are contained in clauses 28A.2 to 28A.7 as follows:

- The provisions of the contract requiring release of retention no longer apply.
- The contractor must provide drawings and other information for the purposes referred to in clause 5.5. This clause is identical to clauses 27.6.1 and 28.4.1 setting out the consequences of determination by employer and contractor respectively.
- Clause 28A.4 is virtually identical to clause 28.4.2 (see section 12.4) requiring removal of equipment from site.
- The employer is obliged to release half the retention held before the date of determination within 28 days of the date of determination. This is subject to any rights which the employer may have before the determination to continue to hold the money. The second half of the retention is to be released as part of the account which the employer must prepare.
- The contractor has two months from the date of determination to supply the employer with all documents necessary for the preparation of a clause 28A.6 account. The employer's obligation to prepare the account with 'reasonable dispatch' is dependent on the contractor complying with his obligation to supply the documents. This is a matter of common sense, but it does no harm for it to be spelled out clearly, because there is a mistaken belief on the part of some parties to a contract that they have no obligations save for those written in the contract.
- The account must contain the following:
  - The total value of work properly executed at the date of determination. The value is to be calculated as though the determination had not occurred. Any other amounts due under the contract must be included; *and*
  - Any sum ascertained for direct loss and/or expense under clauses 26 and 34.3, whether ascertained before or after the determination; *and*

- The reasonable costs incurred by the contractor in removing his plant, etc. from the site; *and*
- The cost of materials properly ordered for the Works for which the contractor has paid or is legally bound to pay. Materials 'properly ordered' are those which it is reasonable that the contractor has ordered at the time of determination; *and*
- If the determination has occurred as a result of suspension following loss or damage to the Works due to specified perils and the cause was negligence or default of the employer or any person for whom he is responsible: any direct loss and/or damage caused to the contractor by the determination.

Money previously paid to the contractor or otherwise to his account must be deducted and the employer has 28 days from the date of submission from the employer to the contractor to pay the balance.

Determination under clause 22C.4.3 gives rise to consequences under clauses 28A.5 and 28A.6 (but not 28A.6.5).

There are distinct similarities between the consequences of all the determination procedures and it should have been relatively simple to have tidied them into one easily digestible set of consequences, albeit with certain exceptions for different determination situations. On the other hand, determination is relatively rare and there may be advantages in being able to relate a set of consequences directly to the particular determination provisions.

It cannot be emphasised too strongly that determination should be treated as a last resort. Even if the determinor successfully negotiates all the pitfalls in the process, the result is likely to be expensive for both parties.

CHAPTER THIRTEEN
DISPUTE RESOLUTION

13.1 *Introduction*

For a hundred years or more, arbitration has been perceived as the ideal procedure for settling construction disputes. It provided a quick and relatively cost effective commercial basis for resolving what were often complex technical disputes. Hence, from the earliest RIBA contracts of the late 1800s through to the mid 1990s editions of their JCT counterparts, all of them, in one form or another and with various degrees of complexity, incorporated provisions and procedures whereby the parties were required to have their disputes determined by an arbitrator rather than by the courts.

Unless the parties agreed to the contrary or there were exceptional reasons to do otherwise, they had no alternative; arbitration was the only formal means available for breaking the deadlock if they were unable or unwilling to compromise. The last decade has seen the advantages of arbitration gradually degraded. Significant increases in the value and complexity of construction claims have brought with them commensurate increases in the time, cost and expertise necessary to resolve such claims. As a result, speedy and inexpensive arbitration is now rather the exception than the rule. There is no denying that there was a time when the potential for incurring significant cost in proceedings may have been a positive incentive for parties to resolve their differences informally and thus to avoid paying lawyers to convert what the parties saw as relatively simple and understandable disputes into ones which were complicated and comprehended only by the respective lawyers and the arbitrator.

It is also true to say that there was a time when few disputants ever saw their arguments carried through to the bitter end. Arbitration was a rarity. But more recently the construction environment has deteriorated significantly. Positive relationships have been replaced by a preoccupation with fault finding and a defensive attitude which ultimately results in a more adversarial and litigious environment. Inevitably more disputes than ever are now running their full course in arbitration. Where previously there may have been reluctance to engage in costly proceedings, now it seems that reluctance has disappeared. Professionals and contractors alike are no longer apprehensive of formal proceedings and even their threat is now often not sufficient to bring parties together.

The deterioration in relationships brought about by those factors reached such proportions that government and industry intervention was inevitable. In 1994 a radical review of existing procurement and contractual arrangements in the UK construction industry was undertaken, out of which evolved *Constructing the Team* (the Latham Report). Latham's recommendation generally received governmental approval in the form of the Housing Grants, Construction and Regeneration Act 1996, commonly known as the Construction Act, Part II of which introduced a contractual system of adjudication.

Except only in contracts for construction of private residences and certain other limited projects specifically excluded by statute, the obligation to incorporate express contractual provisions for adjudication in all other 'construction contracts' is unavoidable. Parties cannot contract out of that requirement. If they either wholly or partly attempt to do so, or if they go further and fail entirely to make express provision in their agreement for adjudication which satisfies the Act, then appropriate terms will be implied. They will then be bound to adopt a statutory scheme for adjudication laid down in subordinate legislation known as the 'Scheme for Construction Contracts (England and Wales) Regulations 1998'. Briefly summarised, the Act imposes obligations on the parties to expressly provide that:

- Either party may give the other notice requiring any dispute or difference between them under the contract to be referred to an adjudicator within seven days of that notice; *and*
- Subject to any post reference agreement between the parties to extend that period, a decision on the matter(s) in issue - and on any necessarily associated matters - must be given by the adjudicator within 28 days of the reference being made; *and*
- The referring party may unilaterally extend the time for the decision to a maximum of 42 days; *and*
- The adjudicator will be given wide powers in relation to the conduct of the proceedings and must be free to take the initiative in ascertaining both the facts and the law within the confines of absolute impartiality.

In connection with design and build projects, those and the other requirements of the Act relating to adjudication are satisfied by use of WCD 98. Amendment, if at all, should be made only after careful and expert consideration and in making any such decision it should be remembered that whatever law it is decided shall be applicable, consideration must be given to what system of law it is that will ultimately have jurisdiction to enforce the parties' respective rights and obligations under that contract. The point is emphasised by a further footnote, (hh to clause 39), reminding the parties that the provisions of the Arbitration Act 1996 do not extend to Scotland. Where the Works are situated in Scotland, forms issued by the Scottish Building Contract Committee containing Scots law and adjudication and arbitration provisions for use under Scots law are recommended for use (see Chapter 14).

Clause 39 deals with the detailed procedures for settlement of disputes. It is divided into three sub-clauses. Clause 39A deals specifically with the commencement and conduct of adjudication under the contract; clause 39B deals with arbitration whilst clause 39C makes provision for the parties to express their preference that litigation be adopted as the final and binding dispute resolution process. In addition, a footnote [dd] at the introduction to clause 39 also alerts the parties to the possibility of them adopting a process of mediation for resolving their differences. However, beyond referring the parties to Practice Note 28 'Mediation on a Building Contract or Sub-Contract Dispute' for more information on that process, no other explanation is given either as to what the process involves, nor how it should be instigated or operated. Why the contract makes brief yet express reference to that option is not, therefore, entirely clear.

13.2 *Adjudication*

13.2.1 Rights to adjudicate

Article 5 of WCD 98, which is introduced into the contract to satisfy section 108(1) and section 108(2)(a) of the Act, provides that:

> 'if *any* dispute or difference arises *under* (the) contract either *Party may* refer *it* to adjudication in accordance with clause 39A [our italics]'

The Act is quite clear; the contract must provide for 'any' dispute to be referrable and on the face of it article 5 clearly satisfies that requirement. Even disputes in relation to the ascertainment of Value Added Tax under clause 3 of the VAT agreement and those arising under clause 31 (Statutory Tax Deduction Scheme) to the extent provided in clause 31.9 are apparently not excluded. However, it should be noted that, both by article 6A and by clause 6 of the VAT agreement, such disputes are expressly beyond the jurisdiction of arbitration under the contract and so, on a plain reading of the contract, the scope and jurisdiction of the adjudication process is, in that regard at least, wider than that of arbitration.

That inconsistency may not have any practical long term consequences and it certainly seems that a party wishing to do so can refer even questions of VAT liability to adjudication under article 5. Nevertheless, it should be realised that to do so is most likely to result in wasted time and cost and may well be inappropriate. Not only is an adjudicator appointed under the contract unlikely to be either able or qualified to decide such matters without at least obtaining the advice of the commissioners, but it must also be remembered that, although not reviewable in arbitration, any decision on the point would be subject to review by the commissioners and then to a further statutory right of appeal in accordance with the contract.

In that case it is thought that it must be in the parties' interests, whether or not they have the right to adjudicate the matter under article 5, instead to refer questions concerning VAT directly in the first instance to the appropriate tribunal, namely the Commissioners of Customs and Excise, as envisaged by clause 6 of the VAT Agreement. Likewise, in the case of other issues reserved to the decision of any other statutory body (such as referred to in clause 31.9, in connection with the Statutory Tax Deduction Scheme), it is suggested that the same principle should apply. Although questions over whether and if so how VAT and statutory tax deduction disputes are to be resolved may generally be of academic rather than practical interest, the same cannot be said for the effects of clause 4.2 of WCD 98.

Article 5 apparently bestows on the parties an absolute right to have any dispute under the contract adjudicated at any time. But, despite the strict provisions of the Act and despite the apparently wide drafting of article 5, clause 4.2 imposes a significant restriction on that otherwise unfettered right. According to clause 4.2, the contractor may ask the employer for his contractual authority for issuing an instruction, in which case the employer must specify in writing which provision he relies on as giving him that authority. If, on receipt of that notice, the contractor then complies with that instruction before either party refers the question for a decision under any of the various dispute resolution procedures available under the contract (including adjudication), then the instruction concerned will be deemed, for all the purposes of the contract, to be authorised by the provision relied on by the employer (see section 5.3.1).

In previous editions, clause 4.2 affected only the parties' right to have the question of the employer's authority to instruct arbitrated. However, the clause is now drafted more widely. Previous reference to arbitration alone has now been replaced with a general reference to 'the procedures under the contract relevant to the resolution of disputes or differences'. Hence adjudication, too, will be time barred unless the employer's reliance on particular provisions is questioned before compliance. Although, at first sight, this amendment effectively imposes limits on the right to refer all disputes to adjudication and so apparently contradicts both the express provisions of the Act and of article 5 of the contract, that is not so. The limitation is in fact merely one of timing. That is to say, disputes about the employer's right or otherwise to issue instructions can undoubtedly still be questioned in adjudication, provided reference to adjudication is commenced before compliance with the instruction concerned. That said, two questions presently remain open. On the one hand it may be argued that even the imposition of such a limitation may, of itself, be contrary to the Act. Section 108(2)(a) provides that the contract must be drafted with the object of enabling either party to refer *any* dispute to adjudication *at any time.* Imposition of any time limits may then be seen as a breach of that statutory requirement, so making clause 4.2 ineffective in so far as it purports to time bar adjudication. Conversely, it may be said with considerable justification that article 5 of the contract does indeed make adjudication possible at any time, since it is only when the contractor takes positive steps to comply with the instruction that that right will be lost. Given that the parties are presumed to know the terms on which they have contracted, it follows that the contractor is free to adjudicate at any time but in complying with the instruction before invoking arbitration he, in effect, elects to give up that right. In short, he could have elected to exercise his right to adjudication at the appropriate time, but chose not to do so.

'either Party'

Albeit that it is likely that the decisions and opinions, etc. of the employer's agent will be called into question, clause 1.3 makes clear that the 'Parties' to the contract are taken to be the employer and the contractor named in the articles of agreement. In that case, although somewhat uniquely under WCD 98 an employer's agent is often appointed (under article 3) 'for the receiving or issuing of ... notices, statements ... or for otherwise acting for the employer under any other of the Conditions', on a strict reading of clause 1.3 it is unlikely that those powers of the agent will extend to commencing or defending adjudication proceedings on the employer's behalf.

Of course, article 3 leaves it open for the employer to limit the agent's authority and it may well be that, for the avoidance of any future doubt, the employer would expressly specify the provisions of article 5 and clause 39A as being matters outside the employer's agent's remit.

'*...either Party* may *refer...*'

Adjudication is not mandatory. If it was intended as such, regrettably neither the legislation (section 108) nor the contract (article 5) say as much. No matter how desirable it would be, as presently drafted there is no reason to presume that adjudication is a prerequisite to commencement of formal arbitration or litigation proceedings. If adjudication is discretionary then it is not difficult to see how certain procedural problems could arise, particularly since both the Act and the contract clearly anticipate that it will be the aggrieved party alone who will wish to take the initiative in electing to refer the issue to adjudication. That, quite possibly, may not be the case.

By way of illustration; where irreconcilable disputes over valuation and payment for changes pursuant to clause 12 of the contract arise, almost invariably it is the contractor who will instigate the dispute resolution procedures under the contract. Despite the option to refer the matter to adjudication, the contractor might (subject of course to any time constraints laid down in the contract) choose instead to refer the matter immediately to arbitration, or litigation as the case may be. On a strict reading of article 5 of the contract and of the Act, the employer may, it seems, elect to invoke his right to have the matter adjudicated. How, if at all, the adjudication would proceed is not clear. The employer who is the respondent in the arbitration will be the claimant (i.e. the referring party) in the adjudication while, likewise, the contractor's role will be reversed. While he would be claimant in the arbitration he would take the role of respondent in the adjudication. Unless one or other action was stayed (postponed) or entirely withdrawn, the situation would become wholly impracticable. Yet, according to both the Act and article 5 of the contract, both parties must have an unfettered right to refer any dispute or difference to adjudication.

As the illustration above shows, that absolute right does not sit well with an adjudication process that is anything other than a prerequisite to the instigation of formal proceedings. It is clearly not sufficient to say simply that the dilemma can be resolved by reading the contract as allowing only the aggrieved party the right to refer the dispute to adjudication, because that too would be contrary to the Act (section 108). Nor is it sufficient to presume that the parties would take the practical steps necessary to resolve the difficulty by agreeing to stay or withdraw one or other of the proceedings. It goes without saying that, even if that could be easily achieved, by the time the matter has escalated to the need for formal dispute resolution it is unlikely that there will be any continuing spirit of co-operation between the parties.

'differences under *the contract'*

Despite the breadth of the adjudicator's powers allowing him to consider *any* dispute or difference including, perhaps, some that are not open to be referred to an arbitrator, those powers are quite sensibly limited in one particular and important respect. Whereas arbitrators have express power to determine all disputes or differences on any matter or thing whatsoever arising under the contract 'or in connection therewith', adjudicators do not. Adjudicators' jurisdiction goes no further than disputes or differences 'under' the contract and so, unlike arbitrators, claims such as those for breach of contract and for general damages are beyond their powers. Similarly, adjudicators have no power to decide issues connected with the contract. Their jurisdiction is therefore, in effect, limited to deciding what are the terms of the contract, interpreting those terms and determining the parties' rights and obligations accordingly.

13.2.2 Notice of intention to refer to adjudication

Once referred to adjudication pursuant to article 5, the adjudication proceedings must then be conducted strictly according to the provisions of clause 39A. Clause 39A.1 puts that beyond doubt. Clause 39A does not dictate in detail how and in what form the 'Notice to

Refer' should be given. Likewise, the Act offers little or no assistance in that regard. Clause 39A.4.1 provides simply that as and when either party requires a dispute or difference to be referred to adjudication, that party shall 'give notice' to the other of his intention to refer the 'dispute or difference, briefly identified in the notice' to adjudication.

'give notice'

Because it is the notice to refer the dispute to adjudication which establishes the future timetable for the ensuing proceedings, it is most important that the first notice is correctly given. Unfortunately, clause 39A.4.1 does not make clear what constitutes effective service. For that, one is, instead, referred to clause 1.5 (see section 2.9). Unlike the provisions of clause 39A.4.2 concerning the subsequent referral of the dispute to the adjudicator, neither clause 1.5 nor clause 39A.4.1 specifies clearly and unequivocally whether the current, and doubtless overused, trend for corresponding by fax, e-mail or other similar instantaneous electronic means constitutes effective service.

Arguably, since clause 1.5 stipulates simply that the recipients must receive notice by way of service to their registered or other agreed address, that implies postal service. But that is by no means made clear and it is at least reasonable to contend that service to one or other of those addresses by fax would likewise suffice. E-mail and other similar electronic communications are, however, quite another matter since in those cases, unlike postal and fax service, the notice (or other message) is, in fact, sent to what can only be described as a holding address, where it is held pending retrieval. The parties can opt for electronic data interchange by clause 1.8, but in annex 2, dealing with that system, it is expressly stated not to apply to notice to be given under the dispute resolution procedures.

Given that the notice to refer is the cornerstone for determining the timetable for the future proceedings, the question whether or not service has been properly effected, it is suggested, should not be left in any doubt and for that reason, if no other, use of facsimile should be avoided. Instead, in strict compliance with clause 1.5, the parties should simply adopt the practice of addressing and sending their notice by pre-paid post, to either the registered or principal address of the opposing company or, where an alternative agreement to that effect has been reached, to any agreed alternative address.

'the dispute or difference, briefly identified'

The precise form and content of the notice to refer is also, according to clause 39A.4.1, something of an open question, except only that the clause requires, as a minimum, that the referring party must briefly identify within it the nature of the matter being referred. Although there may be disadvantages in being unduly explicit and precise at this early stage in the proceedings, it will assist those who may be called on to advise the employer or the contractor if they are at least properly alerted at the outset to the nature and extent of the dispute or difference concerned. The notice should be drafted with that aim in mind.

13.2.3 Appointment of adjudicator

Adjudication is not a process suitable for corporate decision making. Common sense dictates that it is a task properly undertaken only by a single adjudicator acting in his or her individual capacity, and the specific terms by clause 39A.2 of the contract make clear that the adjudicator shall be either an 'individual' agreed by the parties or one who is nominated by the body appointed under the contract to make that nomination. In either event the adjudicator must act in his individual capacity. Thus, where the dispute or difference is one best suited to be decided by the application of particular skill and expertise of, for example, architect, engineer, quantity surveyor, lawyer or the like, the matter cannot be referred to a professional firm or practice. It must, instead, be referred to and decided by a named individual within such a firm or practice or by a sole practitioner engaged in the particular discipline concerned.

Subject only to that constraint, the parties are otherwise free to agree on whom they wish to act. That agreement can be reached as and when the dispute or difference arises or at any time prior to that and it may even be that the parties will make their selection in time to specifically name the individual concerned in the contract. Whenever such agreement is reached and by whatever means, the adjudicator selected must be willing to act in accordance with the standard adjudication agreement for the appointment of an adjudicator – the 'JCT Adjudication Agreement'. He must also be prepared to formally execute that standard agreement (clause 39A.2.1). If, for whatever reason, the parties cannot or do not settle on their choice of adjudicator in advance, or if for some other reason their previous nominee cannot or will not act, then as and when a dispute or difference arises either party may make application for a suitable nomination to be made by the person selected in appendix 1 to the contract as the 'nominator' (clause 39A.2).

'the nominator'

Appendix 1 provides the parties with a choice of four possible 'nominators'. They can and should select one or other of either the President or a Vice President or Chairman or Vice Chairman of the Royal Institute of British Architects or the Royal Institution of Chartered Surveyors or the Construction Confederation or the National Specialist Contractors Council. At the time the contract is concluded the parties should select one nominator from that list. If they do not then by default the task will fall to the President or a Vice President of the Royal Institute of British Architects.

As with any adjudicator agreed on by the parties, a nominee appointed by this process of unilateral application must likewise be willing to signify his agreement to act in accordance with the JCT Adjudication Agreement and must be prepared to formally execute that standard agreement (clause 39A.2.1). Footnote (ee) gives the parties reassurance that whichever nominator is chosen in appendix 1 to fulfil the role, as and when they do so they will nominate as adjudicator only an individual willing to comply with that requirement. For their part, the parties must, jointly with the prospective adjudicator, execute the adjudication agreement. Clause 39A.2.1, and more particularly clause 39A.2.2, has effect to make it a term of the contract that they shall do so and if they do not they will be in breach of contract.

13.2.4 Referral of the dispute

Irrespective of how, or by whom, the nomination and appointment of the adjudicator is made, speed is essential to the successful conduct of the subsequent proceedings. To be effective there must be no scope for either party to stall the process and for that reason the Act lays down a strict timetable within which not only the parties, but the adjudicator too, must deal with the issues concerned.

Both the Act and the contract fall short of specifying a period within which the adjudicator's appointment must be made but they do, however, expressly require that 'any agreement by the parties on the appointment of an adjudicator must be reached with the object of securing the appointment of, *and the referral of the dispute or difference to,* the adjudicator *within 7 days* of the date of the notice of intention to refer (our italics)' (clause 39A.2.2 of the contract and section 108(2)(b) of the Act). That must also be the objective if and to the extent that the appointment is made by way of an application to the nominator (clause 39A.2.2). Perhaps with an abundance of caution, the draftsman of the contract further emphasises the importance of the timetable in the terms of clause 39A.4.1. By that clause, even if the parties should take a full seven days following the notice to refer to reach agreement and to appoint their adjudicator, the party giving the notice to refer must even then, within that same time frame of seven days, also actually refer the dispute or difference to the appointed adjudicator.

Unfortunately, clause 39A.4.1 does not stop there. It goes on to provide for what happens in the event that agreement on, or appointment of, the adjudicator is not achieved within seven days. In that case, according to clause 39A.4.1, the referral must be made immediately following such agreement or appointment. No doubt the practical reasons for this further proviso are easily understood. There is obvious logic in linking the date for referral to the date when the JCT Adjudication Agreement is executed by the parties and by the adjudicator. Yet, on a strict reading it is difficult to see how this extended timing truly satisfies the Act, section 108(2)(b) and section 108(5) of which effectively lay down only one requirement to:

> 'provide a timetable with the object of securing the appointment of the adjudicator and referral of the dispute to him within 7 days of such notice.'

It is therefore difficult to see how a contract drafted to include express provision for dealing with what shall happen if and when the seven day objective is not met can be said to satisfy that objective. However it is looked at, there can be little doubt that, so far as they have the power to do so, the parties must proceed with the aim of achieving referral of the dispute to the adjudicator within seven days from the initial notice to refer and clause 39A.2.2 at least is unequivocal in that regard (see section 2.9 for a discussion on reckoning the period of days).

'referral of the dispute or difference'

As and when the referral is made, both it and any accompanying documents that are sent with it to the adjudicator must simultaneously be copied to the other party (clause 39A.4.2). Original and copy must be served by either actual delivery, special delivery or recorded delivery. Service may also be given by fax but in that case a further copy, which the contract refers to as 'for record purposes', must also be sent, in the words of the contract, 'by first class post or actual delivery'.

It follows from what is said above that, if served by fax, the date (and time) of the referral will then (subject to proof if required) be taken to be when the fax could reasonably have been expected to have been received and read by the intended recipient. Theoretically at least, referral by fax can therefore be instantaneous and can be made any time up until the last hour of the seventh day following notice to refer. That is not a facility that should be relied on when setting one's timetable, bearing in mind that as and when the referral is made it must be accompanied by *all* of the documents relied on.

A dispute involving photographic evidence, drawings and/or many files of papers is quite obviously unsuitable for service by fax and the contract certainly does not sanction referral of some documents by fax and others by post. So any attempt to adopt that procedure would, it is suggested, make use of the fax redundant, thus leaving the date of the referral as the date of receipt (or deemed receipt) of the posted material. Service might then be out of time.

Curiously, so far as the copy for record purposes is concerned, no mention is made of using special or recorded delivery, although clearly either may also be used. It would, it is suggested, be prudent to do so since, as clause 39A.4.2 points out, if sent by special or recorded means there can be no doubt that (subject only to proof to the contrary) the referral, documents or copies, as the case may be, will be deemed to have been received 48 hours after posting, Sundays and Public Holidays excluded. However it is sent, as and when the referral is received by the adjudicator he must immediately notify both parties to that effect (clause 39A.5.1). Unlike the referral, that confirmation is not expressly sanctioned to be given by fax and although that omission may at first sight appear academic, it is useful to note that any response to the referral made pursuant to clause 39A.5.2 must, according to that clause, be made 'within 7 days of the date of referral'. Consequently, where both the referral and the confirmation of its receipt are sent by post instead of by fax, the respondent may in reality have considerably less than that seven days within which to prepare and submit that response.

Whereas no particular form or content is dictated for the notice to refer, the same cannot be said of the referral itself. While the parties are generally free to present their case as they see fit, nevertheless the referral must do so in a way which, as a minimum, clearly sets out:

- Particulars of the dispute or difference; *and*
- A summary statement of the contentions relied on; *and*
- A statement of what remedy or relief is being sought.

Finally, it is of considerable importance that at this stage the referring party must also, according to the contract, include with his referral 'any material he wishes the adjudicator to consider' (clause 39A.4.1). Thus there is no escaping the need for proper forethought and preparation of the case to be put to the adjudicator. Coupled with the obviously short time frame within which the entire proceedings must be concluded, there is, therefore, no room for the copious exchanges of correspondence often associated with litigation or arbitration. Each party's case must be readily understood and from the outset must be properly substantiated by the documents submitted. But there is, without doubt, a fine line between a referral that is sufficiently comprehensive and one that is overly detailed and unnecessarily complex.

In some quarters it has been suggested that a tactical advantage might be gained were a referring party to draft a lengthy, fully particularised and fully argued submission in advance of serving the notice to refer. It is suggested that any such temptation to steal a march on the opposition should be resisted. At first sight it may appear that to do so would effectively extend the contractual seven day time limit for service of the referral. But to do so may ultimately prove a waste of both time and cost.

Clause 39A.5.5 of the contract gives the adjudicator wide powers. He may set his own agenda and while he and the parties must work within the time constraints laid down both in the Act and the contract, the adjudicator nevertheless has absolute discretion to take the initiative in ascertaining the facts and the law in relation to the matter(s) referred to him. Among other things he may require the parties to provide further information beyond that contained in the referral or in the accompanying documentation. Moreover, he may go further and, for example, may:

- apply his own knowledge and/or expertise
- open up, review or revise any opinions, notices, decisions and the like previously given under the contract
- visit the site or any other relevant premises used for the preparation of work in connection with contract and/or call on the

parties to carry out testing or opening up of work, or further testing or further opening up
- take technical or legal advice and/or, after notice to that effect to the Parties, make enquiries of the parties' employees or other representatives.

It is, therefore, quite possible that the adjudicator may stipulate the precise format or even the length of the parties' respective submissions or, alternatively, make some other direction as to the precise form and/or presentation of the referral and response, thus making any advance lengthy drafting abortive. It is not immediately obvious why clause 39A.5.5 of the contract should, even by way of example, go on to explain what might be considered to be within the scope of the adjudicator's wide discretion. Particularly since section 108(2)(f) of the Act provides simply that the contract need only ensure that it makes the general proviso giving the adjudicator absolute power to set his own agenda and allowing him to take the initiative in determining both the facts and the law. It might perhaps have been better if the contract had simply reflected that requirement.

In its present form the contract on the one hand seemingly bestows that wide discretion on the adjudicator, yet on the other it requires that within seven days of the notice to refer and of the referral respectively, the parties must serve on each other a statement of their contentions. In doing so they must also at the same time, according to the contract, assess, assemble and serve all other material on which they will seek to rely in support of their case. It is difficult then to see how that dictate sits comfortably with the adjudicator's absolute discretion (according to clause 39A.5.5) to 'set his own procedure'.

It might be said that on a proper reading of clauses 39A.4.2 and 39A.5.5, only the bare bones of the parties' contentions need be set out both in the referral and in the reply to it made pursuant to clause 39A.5.2, thus leaving the adjudicator free to exercise that discretion. But, if that is indeed the intention then the contract, as presently drafted, is ambiguous on the point. Whereas clause 39A.4.1 expressly uses the term 'summary' in the context of the referring party's submissions, any response that is to be given to it is, according to Clause 39A.5.2, to be in the form of 'a written statement of the contentions' relied on by the respondent. Why the term 'summary' is not used in both clauses is not explained and it is difficult to find any justification for not doing so.

However, on any reading of the contract, one thing is certain.

Much will rest on the quality and content of those respective statements. To find the correct balance between drafting a referral or a response which satisfies the requirements of clauses 39A.4.1 and 39A.5.2 respectively and drafting one that may prove either inadequate or, alternatively, unduly complex will often be a difficult task, particularly given the very short time scale within which the entire exercise must be carried out. Parties should, therefore, give serious consideration from the outset to whether, and if so to what extent, they should take expert advice on the formulation and presentation of their case.

Once the referral is made, the non-referring party (the respondent) has, according to clause 39A.5.2, an express right of reply. If that right is exercised, the reply must likewise be in writing setting out the contentions relied on and must be served complete with all supporting material relied on. Again, too, it must be sent to the adjudicator with a copy sent, simultaneously, to the other party. Post or fax are again the prescribed options for service laid down in clause 39A.5.2.

Proper service of the reply must, according to clause 39A.5.2, be effected 'within 7 days of the date of the referral'. To make proper sense of that proviso, it is suggested that the 'date of the referral' must be taken as the date when the referral was properly served *and* received or at the least it should run from when it would be deemed to have been received, if posted. It is difficult to see how, notwithstanding the strict and limited timetable, time can begin to run otherwise than from the date of actual receipt of the referral. Unfortunately the contract does not expressly put that beyond doubt or, alternatively, stipulate that time for reply should run either from the date of actual receipt by the other side or when receipt by the adjudicator is confirmed or acknowledged by the adjudicator, whichever is the sooner.

It should be noted that a contractual right of reply is not a strict requirement of the Act. The Act requires merely that the referral be made within seven days of notice to refer and says nothing about the non-referring party's rights of reply, let alone when that reply should be given or what it should contain. Without doubt a right of reply must exist. Moreover, as already pointed out above, section 108(2)(f) of the Act provides that the adjudicator has power to take the initiative in setting his own agenda. Hence, even without the express right conferred by clause 39A.5.1, it is inconceivable that the respondent would be prevented from answering the case made by the claimant.

It is essential that the adjudicator not only acts entirely impartially but, most importantly, he or she must be seen to do so, too. That obvious prerequisite is underscored both by the Act (at section 108(2)(e)) and by the contract (at clause 39A5.5) which expressly provides as much. Although not expressed with similar clarity, it is equally important that the adjudicator must ensure that he stays within the strict confines of his jurisdiction. He must not stray beyond deciding only those matters specifically referred to him or necessarily and directly associated with them. Deciding precisely where that jurisdictional line can and should be drawn is often not easy and may on occasions require a narrow view to be taken: *Cameron* v. *John Mowlem* (1990). One aspect of jurisdiction is, however, made clear. On the important question of interest clause 39A.5.8 gives the adjudicator power, subject to any contractual and now to any statutory provision in that regard, to award interest on any capital sum awarded (see The Late Payment of Commercial Debts (Interest) Act 1998). Although the contract is clear in that any such award can only be for simple interest, that power extends to deciding in what circumstances and for what periods a paying party may be liable for such interest.

Subject again to any contractual or statutory impositions to the contrary it also seems open for the adjudicator to decide what rate shall be applied. Whether or not that discretion might be limited to the current statutory rate available in litigation is not made clear. It is submitted that any such limit would not be relevant so far as adjudication is concerned. Any decision in that regard would likely be limited simply by the fundamental requirements of impartiality and natural justice and beyond that, clause 39A.5.5.8 seems to leave no doubt that the matter of the rate of interest is left entirely in the hands of the adjudicator.

13.2.5 Costs

The thorny subject of costs is dealt with succinctly and in such a way as will create a positive incentive for the parties to avoid, or at least informally settle, their differences. When relationships deteriorate and proceedings are threatened, it is the potential costs of those formal proceedings that often plays a significant tactical part in how the matters progress. More often than not it may be that those costs eventually become a positive disincentive to compromising and settling their differences without the need for a formal decision. In the early stages of dispute mania, the threat (and the fear) of incurring - even in the relative short term - substantial costs, often running to tens if not hundreds of thousands of pounds, can be sufficient to force smaller or less wealthy clients, contractors, subcontractors or suppliers to abandon otherwise viable claims or counterclaims. If, on the other hand, matters do proceed formally, it is common for the parties' costs to reach such disproportionate levels in relation to the sums in dispute that the argument becomes one of whether, and if so how much, one party is willing to offer to pay of the other's costs. Those costs then positively militate against the parties being prepared to reach an informal settlement.

It is as much its approach to the question of costs as its use of speed that offers adjudication the greatest chance of achieving its ultimate aim of promoting a quick, economical and if possible informal resolution of genuine disputes. Because clause 39A.5.7 dictates that 'the Parties shall meet their own costs of the adjudication...' irrespective of the outcome, the significant part which costs usually play in other more formal proceedings is largely avoided. Where a party is bound in any case to pay its own costs there is a clear disincentive for generating spurious, or at the very least speculative, claims. Conversely, genuine claims are not stifled by the perceived risk, however small, that the less wealthy party may feel threatened by a potentially large bill for the other's costs. Risk is not altogether obviated. In addition to his otherwise wide powers the adjudicator is also left free (according to clauses 39A.5.7 and 39A.6.1) to decide how his own fees and the costs of any tests or the like carried out under his direction should be apportioned. But even here the contract provides, too, that those costs might be borne in equal proportions irrespective of the outcome and unless the adjudicator specifically directs to the contrary, that will in fact be the default position.

It should be remembered that the parties are, however, and will remain throughout the proceedings, jointly and severally liable for the adjudicator's fees and for all expenses reasonably incurred by him 'pursuant to the adjudication'. That those costs must be incurred 'pursuant to the adjudication' is an important and useful safeguard and parties faced with a significant bill from the adjudicator should not be reluctant to ask for justification both that the costs are reasonable and that they have reasonably been incurred in the furtherance of the adjudication.

13.2.6 Effect of the decision

Once made, the decision of the adjudicator is binding on the parties until the dispute or difference is finally determined by arbitration or by legal proceedings or by an 'agreement in writing between the parties' made 'after the decision of the adjudicator has been given'.

'the dispute or difference'

In keeping with the philosophy of adjudication generally, the contract makes clear that any subsequent referral of the dispute either to arbitration or to litigation will not amount to an appeal against the adjudicator's decision. It will be a complete retrial of the matter concerned, as if no decision had been made by the adjudicator (footnote (gg) to clause 39A.7.1). If arbitrated, the issues will be dealt with according to the rights and procedures laid down in the Arbitration Act and subject to any agreed arbitration rules. If litigated, the parties will follow the full judicial process. Whether revisited in arbitration or in litigation it is important to note that it is 'the' dispute (i.e. the same dispute or difference as that decided by the adjudicator) which must be revisited. The importance of this distinction so far as it applies in arbitration is perhaps best demonstrated when considering clause 39A.7.1 in the context of section 14 of the Arbitration Act 1996, discussed more fully below.

'agreement in writing between the Parties'

Except only in certain limited circumstances, contracting parties generally are free to agree additions to and/or variations of the terms of their existing contract as and when they wish. Although the advantages of written evidence of such a variation or addition are obvious, with few exceptions such as contracts executed as a deed or those where statute dictates otherwise (for example, the Law of Property Act 1925), the parties may also choose whether to make their variation or addition in writing or orally. Although those general rules are largely unaffected by WCD 98, clause 39A.7.1 expressly requires that any agreement as to the binding nature of the adjudicator's decision must be made in writing. That is a sensible precaution.

No guidance is given concerning the form or the content of such written agreement and so it may be that it could be effected by a simple exchange of letters; one party writing to the effect that it is willing to be bound by the decision, the other replying likewise stating a willingness to be bound. However, it is submitted that, although that may in many circumstances amount to an agreement *evidenced* in writing, clause 39A.7.1 implies that something more specific is required. Read strictly, it appears that the agreement must be written and not merely evidenced in writing. That suggests that any such agreement should, and for the avoidance of doubt must, be brought together in one document, signed and countersigned by the parties.

'after the decision of the adjudicator has been given'

This further significant safeguard against abuse of the process has much to commend it. Courts have been slow to give legal effect to the concept of inequality of bargaining strength at the time of concluding contracts, but in reality there can be little doubt that parties to a contract rarely, if ever, have equal standing when the contract is drawn up and agreement to it is concluded. The commercial reality is that contractors need continuity of work as do subcontractors and without that safeguard there must be the real prospect of the dominant party, be it employer to contractor or contractor to subcontractor, imposing onerous provisions by means of clever drafting, precluding any right to have any future adjudicator's decision revisited. No doubt it is for that reason that the Act and the contract expressly preclude any such possibility by requiring that agreements to be bound by the adjudicator's decision can be made only after the 'giving' of the decision concerned, thereby removing any imbalance that might otherwise exist between the parties when first they contracted.

Unless and until agreement is reached to be bound or until the adjudicator's decision is ratified or reversed in arbitration or litigation, the parties are bound by it. They must give effect to it (clause 39A.7.2) and if they fail to do so legal proceedings may be begun in order to secure such compliance (clause 39A.7.3), even where all disputes and differences are agreed under the contract to be referred to arbitration as opposed to litigation: *Macob Civil Engineering* v. *Morrison Construction* (1999).

As a general rule, where parties have settled on arbitration as their process of last resort for the settlement of disputes or differences, the courts will have no jurisdiction, except in the case of certain very limited rights of appeal, to consider or enforce the parties' rights. Clause 39A.7.3 and article 6A alter that position so that the enforcement of an adjudicator's decision remains a matter for the courts. By clause 39A.7.3:

> 'If either Party does not comply with...'

And by article 6A:

'...any dispute or difference ... except in connection with the enforcement of any decision of an adjudicator ... shall be referred to arbitration...'

It is worth noting that, subject only to any act of bad faith either by the adjudicator or by anyone acting as his employee or agent, the adjudicator and such employees or agents enjoy immunity from liability for anything done or not done in the discharge or purported discharge of his functions as adjudicator (clause 39A.8). If for no other reason, the parties should give careful thought to the suitability of any candidate that they consider nominating to act.

13.3 *Arbitration*

13.3.1 General

After the decision in *Beaufort Developments (NI) Ltd* v. *Gilbert-Ash NI Limited* (1998) it remains to be seen whether the previous long standing use of arbitration will continue as the usual method by which construction disputes will be finally resolved. For some 14 years, between 1984 and 1998 when the House of Lords gave judgment in the *Beaufort* case, the courts' power and jurisdiction to open up, review and revise architects' certificates and opinions given under JCT contracts was severely curtailed. Until the *Beaufort* judgment, the leading case of *Northern Regional Health Authority* v. *Derek Crouch Construction Co Ltd* (1984) held sway with the effect of preventing the courts from opening up, reviewing or revising architects' opinions and certificates issued under the JCT family of contracts. Since construction disputes invariably raise questions about the correctness or otherwise of such certificates and opinions, the *Crouch* decision was perhaps the single most important factor influencing parties in their decision as to whether to adopt arbitration or litigation as their preferred method of dispute resolution.

Crouch effectively gave the parties no realistic alternative. Arbitration was without doubt the most appropriate option. But after *Crouch* was overturned and since the courts are no longer constrained in that way, it remains to be seen whether contracting parties (or more likely their legal advisors) will continue to favour arbitration. Signs are that they will not and one must wonder whether the express incorporation of litigation as an alternative in the WCD 98 (at appendix 1, article 6B and at clause 39C), together with specific reference in the contract to guidance note WCD

Amendment 12 offering a resumé of some of the more important factors to be taken into account in arriving at that decision, might be seen as signifying the beginning of a trend away from arbitration.

Parties wishing to adopt litigation rather than arbitration must ensure that they complete the appendix correctly to reflect properly that intention. They must amend the standard form contract by deleting the reference to clause 39B in appendix 1, for if they do not do so, by default, the arbitration agreement will take effect. Subject only to the very limited exceptions specified in article 6A, all other disputes and differences must then be referred for a final and binding decision to arbitration. Not to do so will invariably result in proceedings being begun in the wrong forum, so making the claimant liable not only for their own but also the other party's costs of rectifying that mistake irrespective of the final outcome of the subsequent arbitration (Arbitration Act 1996 section 9).

Unlike a number of the other JCT contracts, WCD 98 puts no restrictions on when certain matters may be referred to arbitration and under it arbitration can take place on any matter at any time.

Arbitrators appointed under a JCT arbitration agreement such as the one in WCD 98 were previously, and still are, given extremely wide express powers. Their jurisdiction is to decide any dispute or difference arising under the contract or connected with it (article 6A). That general authority is extensive (*Ashville Investments Ltd* v. *Elmer Contractors Ltd* (1987)) and by clause 39B.2 extends to:

- Rectification of the contract
- Directing the taking of measurements or the undertaking of such valuations as he thinks appropriate
- Ascertaining and awarding any sum that he considers ought to have been included in any payment
- Opening up, reviewing and revising any account, opinion, decision, requirement or notice issued, given or made and to determine all matters in dispute as if no such account, opinion, decision, requirement or notice had been issued, given or made.

Prior to enactment of the Arbitration Act 1996 and before the advent of WCD 98, unless otherwise agreed the conduct of arbitrations instigated under the JCT contracts was governed by the 1988 JCT Arbitration Rules. Those rules offered a choice of procedures, common to all of which was the implementation of a strict pre-agreed timetable for the proceedings. Failure to meet those timetables could give rise to important sanctions. Following the 1996 enactment and with the introduction of WCD 98, a number of significant changes have been made to the way in which the arbitration process is to be conducted. No longer will the 1988 JCT Rules apply. New rules, the Construction Industry Model Arbitration Rules (CIMAR) 1998 edition and current at the contractual base date, now govern the proceedings (clause 39B). Those rules, coupled with the extensive revisions to the arbitration provisions in clause 39B of the contract, now amount to a fundamental overhauling of the arbitration process necessary to bring it into line with the new 1996 Arbitration Act.

Provisions relating to arbitration now first appear in article 6A. Subject to the exercise of any prior right, conferred by article 5, to have the issue initially adjudicated, all disputes or differences arising under or connected with the contract and arising between contractor and employer, or the architect on his behalf, 'shall' be referred to arbitration. If either party, mistakenly or otherwise, attempts to bypass the agreed route to arbitration and instead begins proceedings in the courts, they will very soon come unstuck. There are three exceptions to that position, the following matters being specifically excluded from the arbitral process:

- Disputes about Value Added Tax (supplemental condition A7)
- Disputes under the statutory tax deduction scheme, provided statute dictates some other method of resolving the dispute (supplemental condition B8)
- Matters in connection with the 'enforcement' of any decision of an adjudicator.

Where the issue is one that falls under one or other of the first two exceptions, the appropriate statutory tribunal, in the former case the Commissioners for Customs and Excise, will have authority to hear and decide the matter. Only when the question is one concerning non-compliance with any decision previously made by an adjudicator will the parties be free to begin legal proceedings to secure such compliance. In such cases, the signs now are that the courts will take a robust view of the parties' obligation to conform with any such decision. Even then, however, the courts will still only play an interim role. They may make an Order regarding the enforcement of the adjudicator's decision but only in so far as that Order will be made pending a final determination in arbitration of the adjudicated matter: *Macob Civil Engineering* v. *Morrison Construction* (1999).

With the incorporation of clauses 39B.4 and 39B.4.1 into their agreement, the employer and contractor agree, pursuant to section

45 of the Act, that either party may by proper notice to the other and to the arbitrator apply to the courts to determine any question of law arising in the course of the reference. Although not now the first and only available means of formal dispute resolution, arbitration, if chosen in preference to litigation, will remain for all material purposes the last resort. As such the parties and architect alike should do all they can to avoid it. It is like marriage; it should not be entered into lightly or unadvisedly.

Even with the advantages of the previous JCT Arbitration Rules and now with their more recent counterpart, CIMAR, arbitration can be a costly, time consuming and inevitably risky venture. The eventual outcome is always uncertain and those involved in the contract should do everything possible to avoid it. Some contractors nevertheless will threaten arbitration over trivial matters in an attempt to persuade the architect to alter a decision which they dislike. Others and similar-minded employers alike may use the risk, however small, of a successful outcome with an associated award for costs in their favour to force an offer of settlement against what might otherwise be considered merely a speculative and unmeritorious claim or set-off. Unfortunately, even with the recent review of dispute resolution procedures and introduction of the adjudication process, that approach is unlikely to disappear overnight. Wise contract administrators must therefore deal with speculative threats of arbitration firmly. Despite even the most strenuous efforts to do so, it will not always be possible to avoid arbitration and so employers and contractors must ensure that they properly appreciate how the process operates. Only then can they recognise the possible consequences of embarking on formal proceedings, both in terms of time and cost.

It is commonly misunderstood that the arbitration process is nothing more nor less than an airing of each party's opinions and arguments in a semi-formal debate during which each party simply argues out their position on a rather ad hoc basis, before the arbitrator then decides, in a rather casual manner, whose story he prefers. Indeed, it is not uncommon for contractors and architects alike to expect the arbitrator simply to 'split the difference' where the dispute is one over the valuation of variations. This is not the case. The parties should not confuse the difference in form and style with informality. Arbitration is a variant of, and nothing short of, formal legal proceedings.

Whether or not informality and an inquisitorial approach by the arbitrator would prove a more satisfactory approach is an open question, but that is not presently the case. Except in certain limited circumstances the arbitrator will not test the parties' veracity by his own direct questioning and intervention. Nor will he invite comment and/or response from either side. Present day arbitration, at least so far as the construction industry is concerned, is a far cry from that rather informal and inquisitorial approach. Like the judicial system, arbitration is adversarial albeit that it offers a degree of flexibility not so far available to litigants.

Employer and contractor are not only free to agree who should be appointed as, or who should appoint, the arbitrator, they also have freedom to agree important matters such as the form and timetable of the proceedings. This raises the possibility of a quicker procedure than would be the case in litigation and even matters such as the venue for any future hearing might be arranged to suit the convenience of the parties and their witnesses.

Confidentiality, too, is another important aspect of arbitration. Hearings are conducted in private, not in an open court. Parties are free to choose whether to represent themselves or whether to be represented and by whom. They need not, in the traditional way, be represented by solicitor and counsel. They may choose to represent themselves or be represented by what has commonly been referred to as 'para legals' with expertise and qualifications in one or more construction profession coupled with legal qualification and experience in the care and conduct of such proceedings.

13.3.2 Rules of procedure

When the JCT Arbitration Rules were introduced in 1988, the industry had a tailor-made procedure for conducting and settling its disputes in arbitrations. The JCT Rules, as they then were, had their critics, no doubt with some justification. However, they offered parties a reasonable and more importantly pre-agreed framework according to which they were bound to prepare and present their respective cases. To that extent at least they were of real benefit. They removed the opportunity for time wasting contentious debate over purely procedural matters. Moreover, the rules were agreed at the time the contract was concluded and not when disputes or differences had arisen and relationships deteriorated to the point where any such agreements would become less likely.

Introduction of the Arbitration Act 1996 brought with it the need for a radical rethink and a different approach. After extensive consultation with interested bodies over some 18 months or more, the JCT Rules have now given way to the Construction Industry

Model Arbitration Rules. Primarily, CIMAR are aimed at providing parties with a more appropriate and user friendly framework, briefly yet clearly supplementing the specific powers and duties of the arbitrator and of the parties as now defined under the Act. They provide clear guidance to users and arbitrators alike involved in arbitration under the JCT contracts.

Parties are now not specifically warned against the strictures of the timetables set by the new rules nor are they invited to consider whether or not to adopt them. Indeed, for the parties to avoid the application of CIMAR would require amendment to article 6A along with wholesale amendment to clause 39B. More to the point, such amendment would no doubt be largely fruitless since much of what now appears in CIMAR, and in the contract, merely reflects provisions of the Arbitration Act with which the parties and the arbitrator must comply. Furthermore, any amendment to WCD 98 should not be undertaken lightly. Only after careful and expert consideration should such alterations be made and in the case of CIMAR it is difficult to think of any good reason why the parties contracting in England and under English law should want to avoid their use.

For those contracting under Scottish Law the position is somewhat different. The provisions of the Arbitration Act 1996 do not extend to Scotland and parties are reminded, by a footnote (hh) to clause 39B.5, that where the site of the Works is situated in Scotland the forms issued by the Scottish Building Contract Committee should be used since they contain arbitration and adjudication provisions specifically appropriate for use where the proper law of the contract is Scots law. Guidance concerning those alternative provisions is issued by the Scottish Building Contract Committee (SBCC).

Arbitrations begun under contracts made using WCD 98 and subject to the law and jurisdiction of the English courts according to article 6A and clause 39B must be conducted subject to and in accordance with the 1998 edition of CIMAR, current at the base date stated in the appendix. Notably, if any amendments have been made to those rules since that base date then the parties may jointly agree to instruct the arbitrator, in writing, to conduct the reference according to that more recent version (clause 39B.6). In addition to their express agreement to use CIMAR, the parties also expressly agree that the provisions of the Arbitration Act 1996 shall apply too (clause 39B.5), irrespective of where the arbitration or any part of it will be conducted. It should be noted that this reference is to the locality of the proceedings and does not affect the situation concerning the relevance of the location of the site as outlined in footnote (hh) referred to above.

13.3.3 Conduct of proceedings

The new rules have much to commend them. They continue to offer the parties a choice of three broad procedures by which to conduct the proceedings: documents only, short hearing, and full procedure.

Documents only procedure

Experience of disputes that have commonly arisen under CD 81 leads to the view that the documents only procedure will only rarely be a viable option. Nevertheless, it is a much maligned and often ignored option that on occasion can offer real economies of time and cost. It is best suited to disputes capable of being dealt with in the absence of oral evidence and where the sums in issue are modest and do not warrant the time and associated additional expense of a hearing. Parties, either simultaneously or sequentially as the arbitrator directs, will serve on each other and on the arbitrator a written statement of case which, as a minimum, will include:

- An account of the relevant facts and opinions relied on
- A statement of precisely what relief or remedy is sought.

If factual evidence of witnesses is to be relied on, witness statements (or 'proofs'), signed or otherwise confirmed by the witnesses concerned, will also be included with the statement of case. Similarly, if the opinion of an expert or experts will be relied on those, too, will be given in writing, signed (or otherwise confirmed) and incorporated. There will be a right of reply and if any counterclaim is made that, too, may be replied to before the arbitrator, if he wishes, puts questions or asks for further statements as he considers necessary or appropriate. Should he ultimately wish to do so the arbitrator may, and the rules provide that he can, set aside a day or less during which to question the parties and/or their witnesses. If that does in fact happen then the parties will have an opportunity to comment on any additional information that may then have become known.

Given the type and size of issues most commonly suited to this type of procedure, more often than not the arbitrator will be in a position to reach his decision within a month or so of final exchanges and questioning.

Short hearing procedure

Although the short hearing is another unlikely option given the nature and complexity of disputes common to design and build contracts, it is nevertheless a useful procedure which limits the time that the parties have within which to address the matters in dispute orally before the arbitrator. That time can, of course, be extended by mutual consent but without that agreement no more than one day will be allowed during which both parties will have a reasonable opportunity to be heard. Before then, either by simultaneous exchange or by way of consecutive submissions, each party will provide to the arbitrator and to each other a written statement of their claim, defence and counterclaim (if any), as the case may be. Each such statement will be accompanied by all documents and any witness statements that it is proposed to rely on and where appropriate, either before or after the short hearing, the arbitrator will have the opportunity to inspect the subject matter of the dispute should he wish to do so. This is a procedure particularly well suited to issues which might readily be decided principally by such an inspection of work, materials, plant and/or equipment or the like and is useful, too, where the arbitrator generally can decide the issues and make his award within a short time frame of around a month or so after considering the statements and hearing the parties.

In appropriate circumstances expert evidence can be presented by one or more usually both parties. However, that can often be costly and often is unnecessary, particularly where the arbitrator has been chosen or appointed specifically with his own specialist knowledge and expertise in mind. In that case, parties can quite readily agree to allow the arbitrator to use that specialist expertise when reaching his decision and so the use of independent expert evidence under the short procedure is all but actively discouraged under rule 7.5 of CIMAR which precludes any party calling such expert evidence from recovering the costs of doing so, except where the arbitrator determines that such evidence was 'necessary for coming to his decision'.

Subject to the reservation expressed at the introduction above, this short procedure with a hearing has many tangible advantages over the full blown and often extremely expensive procedure with a full hearing (discussed below). It is ideally suited to many common small and not unduly complex disputes and provides for a quick award with minimum delay and associated unnecessary cost. In practice there is nothing to prevent either the employer or contractor suggesting this procedure and a wise arbitrator should give its merits serious consideration.

Full procedure

Where neither of the preceding shorter options is deemed satisfactory, or where, for example, there is significant disagreement over the essential facts or technical opinion evidence then the rules, and in particular rule 9 of CIMAR, make provision for the parties to conduct their respective cases in a manner analogous to conventional High Court proceedings and offering the opportunity to hear and cross examine factual and expert witnesses.

Like litigation, arbitration must offer finality in deciding not only simple but also the most complex of disputes perhaps involving hundreds of thousands or even millions of pounds. In such cases the contract and the rules must cater for the whole range of issues that might arise and must offer a workable framework around which the proceedings should be conducted. Consequently they must be capable of modification in the same way as they are in the High Court, so that they can be properly and cost effectively tailored to the particular complexities of any given disputes. For that reason the rules make clear that the arbitrator is free to:

> 'permit or direct the parties at any stage to amend, expand, summarise or reproduce in some other format any statements of claim or defence so as to identify the matters essentially in dispute including preparing a list of matters in issue.'

Subject to any such directive, parties operating under the full procedure will exchange formal statements comprising, in extreme cases, claim, defence and counterclaim (if any), reply to defence, defence to counterclaim and reply to defence to counterclaim. Each must be sufficiently particularised to enable the other party to answer each allegation made, and must as a minimum set out:

- The facts and matters of opinion which will be established by evidence and may include statements concerning any relevant point(s) of law
- Evidence, or reference to the evidence it is proposed will be presented, if this will assist in defining the issues
- A clear statement of the relief or remedies sought such as, for example, the specific monetary losses claimed set out in such a

way as will enable the other party to answer or admit the claim made.

The arbitrator should, and no doubt will, give detailed directions concerning both the timing for service of those respective statements and all other steps necessary, or likely to be necessary, in the period leading up to the hearing. Commonly those directions will include orders regarding the time within which either party may request further and better details of their opponent's cases and the reply to any such request. Directions may also be given requiring the disclosure of any documents or other relevant material which is or has been in each party's possession. More likely than not, the parties will be required to exchange written statements setting out any evidence that may be relied on from witnesses of fact in advance of the hearing. If expert evidence is also being relied on, the timing for preparation and exchange of written statements from those, too, will be the subject of a direction from the arbitrator.

13.3.4 Appointing an arbitrator

Subject to any adjudication that is already underway, under WCD 98 the parties need not wait until completion of the Works, or until determination or alleged determination of the contractor's employment under the contract, before proceeding to arbitration. Provided a 'dispute or difference as to the construction of this contract or any matter or thing of whatsoever nature arising thereunder or in connection therewith' exists between the parties, either party can begin arbitration proceedings. The first step in the procedure is for one party to write to the other requesting concurrence in the appointment of an arbitrator. In most cases, it will be the contractor who does so but that need not be the case. There is no reason why the employer should not take the initiative and whoever does so, the proceedings are formally begun when one or other party serves on the other a written notice, in the manner provided by rule 2.1 of CIMAR, 'identifying the dispute and requiring agreement to the appointment of an arbitrator'.

When inviting agreement to the appointment of an arbitrator the party serving the notice should name at least one person that he proposes should act as arbitrator (CIMAR rule 2.2). If possible the names of up to three 'individuals' considered suitably qualified to act should be suggested (clause 39B.1.1). Beyond the obvious requirements that nominees be competent, experienced and suitably qualified they must also be independent. They must have no live connection with, or interest in, either of the parties. Nor should they have connections with any matter associated with the dispute. In choosing a suitable arbitrator it is essential for all concerned to have a proper understanding of the nature of the dispute and of the sums involved and so the parties must ensure that these are clearly appreciated from the outset. The Chartered Institute of Arbitrators, International Arbitration Centre, 24 Angel Gate, City Road, London EC1V 2RS (Telephone 0171 837 4483) is always willing to suggest the names of suitable people on request.

However he is appointed, the arbitrator must be:

- Independent
- Impartial, with no existing relationships with either the employer, contractor or anyone else involved or associated with the parties or the issue to be decided
- Technically and/or legally qualified, as appropriate.

Wherever possible and even at the risk of swallowing some pride, it is sensible for the parties to make every effort to agree on a suitable candidate. In the nature of things, because they are already in dispute, it will often not be possible and the parties will be unwilling or unable to do so. Indeed, requests for agreement on the appointment of an arbitrator are often even ignored entirely or rejected without thought. In that case, deadlock is avoided by clause 39B1.1 and by the provisions of rule 2.3 of CIMAR, both of which require that if the parties cannot agree on a suitable appointment within 14 days of a notice to concur (or within any agreed extension to that period), an arbitrator will be appointed by a third party. In a similar way to adjudication, when first completing the appendix, the parties are given the choice of appointing bodies. All but one of those listed should be deleted thus leaving the agreed appointer as either the President or a Vice-President of the Royal Institute of British Architects, or the Royal Institution of Chartered Surveyors, or the Chartered Institute of Arbitrators. Should no appropriate deletion be made or should no agreement be reached at the time of contracting, the task will fall, by default, to the President or a Vice-President of the Chartered Institute of Arbitrators. He will make a suitable appointment after written application by either party, provided, of course, the time for reaching agreement has expired without success.

It will be remembered that the arbitral proceedings are begun in respect of 'a dispute' when one party serves on the other 'a written notice of arbitration identifying the dispute' and requiring him to agree to the appointment of an arbitrator. Those words are of considerable importance. Notice of arbitration in connection with a particular (and often long running) dispute commonly promotes a counterclaim from the recipient of the notice, the respondent. The words raise important questions about whether, and if so how, any such counterclaim might be brought within the jurisdiction of the arbitration begun by the initial notice.

It has long been custom and practice in construction disputes for respondents simply to raise their counterclaim formally at or about the time of serving their defence to the primary claim. That custom and practice now seems to be in jeopardy since it may be attacked by a claimant wishing, albeit as a short term strategy, to frustrate the respondents' attempts to automatically bring that counterclaim into the proceedings. The significance of the point is emphasised further by rule 3.6 of CIMAR which makes clear that: 'arbitral proceedings in respect of any other dispute are begun when notice of arbitration for that other dispute is served'.

Such tactics are more than merely academic point scoring and although a claimant's insistence that the respondent serve a fresh notice in respect of his counterclaim is generally nothing more than a temporary hindrance, clause 39B.1.1 and rule 2.3 raise very real practical issues. Doubts over whether, and if so when, a counterclaim has properly been brought within the jurisdiction of the original arbitration may well have effect later on both the existence and the extent to which either party gains protection from liability for costs where previous without prejudice offers of settlement may have been made. Moreover, as rule 3.6 emphasises, it is of considerable practical importance if it is only long after the initial arbitration has been commenced that the respondent or his representatives then realise that a fresh notice, and hence new proceedings, are necessary if a counterclaim is to be pursued. At its most extreme, the counterclaim might even then be time barred if the realisation dawns only after any contractual or statutory time limit for the commencement of proceedings has expired (see section 10.6).

Notice to concur in connection with a previously undisclosed counterclaim may also be susceptible to attack as invalid simply on the premise that, at the time the notice was given, no 'dispute' then existed. Quite simply, it may be argued that on a strict view no opportunity has been given for the subject matter of the counterclaim to be considered and rejected or even possibly accepted, and so no 'dispute or difference' can yet be said to exist: *Hayter* v. *Nelson* (1990).

Subject to any time bar such as that mentioned above, it seems clear that further notices to concur can be served either before an arbitrator is initially appointed (CIMAR Rule 3.2) or after his appointment (CIMAR rule 3.2 and clause 39B.1.3 of WCD 98) and in the latter case CIMAR rule 3.3 will apply to the subject matter of that subsequent notice.

When determining which rules apply to the service of any further notice to concur, establishing the precise timing of the initial appointment of the arbitrator may well be important. It should be remembered that, if the arbitrator's appointment is made by agreement, it will take effect when he confirms his willingness to act, irrespective of whether by then his terms have been agreed. If, on the other hand, his appointment is the result of an application to the Chartered Institute of Arbitrators, or other agreed appointing body, it becomes effective, whether or not terms have been agreed, when the appointment is made by the relevant body (clause 39B.1.1 and CIMAR rule 2.5). Arbitrators are professional people and charge professional fees. There is no fixed scale of charges for their services and an individual's fees will depend on their experience, expertise and often on the complexity of the dispute; so it should be stressed again that the timing of the appointment is not linked to any agreement to meet those terms.

Given the scope for arguments of the kind outlined above, respondents receiving a notice to concur should, it is suggested, waste no time in taking proper expert advice on how best to respond. Claimants and counterclaimants would also be well advised to ensure strict compliance with the rules and should take proper advice both as to the timing and the content of any notice which they either intend to serve or have received.

Once appointed, the arbitrator will consider which of the procedures summarised above appears to him (and to the parties) to be most appropriate in terms of not only the size and complexity of the particular dispute but also its suitability as a forum for the parties to put their own case and to answer their opponent's case. At the same time the dispute must be kept in context. The arbitrator must have an eye to the format that will best avoid undue cost and delay and for even the most experienced arbitrator that is often a most difficult balancing act. Parties should, therefore, as soon as possible after his appointment, provide the arbitrator with an outline of their disputes and of the sums in issue along with an indication of which procedure they consider best suited to them.

After due consideration of both his own and the parties' views he will generally arrange a meeting at which the parties or their representatives will attend before him to agree or decide on which procedure shall apply. Any directions concerning particular tailor-made procedures that may be required to suit the specific circumstances will also be decided at that time. For example, specific time constraints may be imposed, or directions regarding the early hearing of certain questions of liability, or other preliminary matters, may be given. Although parties are always free to conduct their own case, if disputes have reached the stage of formal proceedings it is at this stage often better to hand over care and conduct of the proceedings to solicitors or construction consultants with particular expertise in the care and conduct of arbitration proceedings.

13.3.5 Arbitrator's powers

Under the 1950 and 1979 Arbitration Acts, arbitrators already had extremely wide powers. With enactment of the 1996 Act those powers have not diminished and in numerous important respects have now grown significantly wider.

Further underpinning of the arbitrator's wide jurisdiction and discretion is given by the contract (clause 39B.2) which provides, subject to article 6A and to clause 30.8.1, that without prejudice to the generality of his powers the arbitrator has power to:

- Rectify the contract so that it accurately reflects the true agreement made by the parties, i.e. to order the correction of errors if the contract fails to represent what the parties actually agreed. It must be shown that the parties were in complete agreement on the terms of the contract, but by an error wrote them down wrongly.
- Direct such measurements and valuations as may in his opinion be desirable in order to determine the rights of the parties.
- Ascertain and award any sum which ought to have been included in any certificate.
- Open up, review and revise any certificate, opinion, decision, requirement or notice.
- Determine all matters in dispute which shall be submitted to him in the same manner as if no such certificate, opinion, decision, requirement or notice had been given.

The last two powers are most important. He can review the employer's decisions and opinions and can, in effect, substitute his own opinion for that previously formed by the employer. This is especially important in the context of matters such as extensions of time and the revaluation and payment of claims made by the contractor for direct loss and/or expense.

Under the contract, four matters are expressly excluded from this otherwise wide power:

(1) Clause 4.2: where the contractor has asked the employer for his contractual authority for issuing an instruction and the employer has responded, unless the matter is submitted to immediate arbitration, the employer's answer is 'deemed for all the purposes of this contract to have been empowered' by the provision specified.
(2) Clause 22C.4.3.1: where a party determines the contractor's employment within 28 days of loss or damage the party in receipt of the notice has just seven days in which to invoke the dispute resolution procedures.
(3) Clause 30.5.5 or clause 30.5.8: the Final Account and Final Statement submitted by the contractor or by the employer as the case may be will be conclusive as to the balance due between the parties except only if and to the extent that the employer or contractor disputes anything contained in such Final Account and Final Statement.
(4) Clause 38.4: the amount of fluctuations.

Beyond those and any other specifically agreed limits to his powers provided for in the contract, the arbitrator's statutory powers are considerable and it should be realised that only in the most limited circumstances can that power be revoked unless the parties have already agreed in what circumstances that may be done, or otherwise act jointly in writing to do so.

By way of example, an arbitrator may:

- Order whether and if so which documents or classes of documents should be disclosed between and produced by the parties: section 34(2)(d)
- Decide whether and to what extent he should take the initiative in ascertaining the facts and the law: section 34(2)(g)
- Order whether the strict rules of evidence, or any other rules, as to the admissibility, relevance or weight of any oral, written or other material which either party may wish to produce shall apply: section 34(2)(f)
- In appropriate circumstances dismiss the claim or otherwise continue the proceedings and make an award based on the

information before him in the absence of a party or without any submissions on his behalf in the event of inexcusable delay or other failure to properly and expeditiously conduct the arbitration: section 41(1) to section 41(4)
- Make more than one award at different times on different aspects of the matters to be determined: section 47(1)
- Direct (and timeously review any previous direction) that the recoverable costs of the arbitration, or any part of the arbitral proceedings, shall be limited to a specified amount: section 65(1) and section 65(2)
- Make an award allocating liability for the recoverable costs of the arbitration as between the parties: section 61(1) and section 61(2)
- Award interest: section 49(1) to section 49(6)
- Take legal or technical assistance or advice: section 37
- Order security for costs: section 38
- Give directions regarding the inspection, photographing, preservation, custody or detention of in relation to any property owned by or in the possession of any party to the proceedings which is the subject of the proceedings: section 38.

13.3.6 Awards and appeals to the High Court

Arbitrators' awards are binding except only in the most limited circumstances (clause 39B.3). Under section 45 and section 69 of the 1996 Act, unless the parties otherwise agree, they may, subject to certain time limits and other procedural requirements, appeal to the courts on questions of law. Under section 45, subject to any agreement to the contrary and subject to proper notice, they may apply to the courts to have determined any question of law arising during the course of the reference.

Clauses 39B.4 and 39B.4.2 of the contract put beyond doubt the parties' right to appeal to the courts on certain limited issues. Both the employer and contractor will thereby have expressly agreed that, pursuant to section 69 of the Arbitration Act 1996, either party may appeal to the courts on any question of law arising out of an award made in an arbitration under the contract. Similarly, by virtue of clause 39B.4.1 of WCD 98 they will have agreed, pursuant to section 45 of the Act, that on written notice to the arbitrator and to the other party, either the employer or contractor may apply to the courts to determine any question of law arising in the course of the reference.

If they agree, after the arbitration process has commenced, the parties may exclude those rights of appeal but to do so would be a serious step. It should be noted that, even an agreement simply to dispense with a reasoned award from the arbitrator, if properly made, may also have effect to exclude that existing right of appeal. Quite obviously, entering into any such exclusion agreement is a very serious step and should not be undertaken without careful expert legal advice.

13.3.7 Third party procedure

Clause 39B.1.2 combined with rules 2.6, 2.7 and 2.8 of the CIMAR are an attempt to introduce into the arbitration a type of third party procedure, akin to that available in litigation, by facilitating two or more related arbitral proceedings in respect of the Works being heard and decided by one arbitrator. The intention of clause 39B.1.2 is no doubt to make provision so that all parties will join in the arbitration if the dispute raises issues which are substantially the same as or connected with issues raised in a related subcontract dispute either already referred or about to be referred to arbitration. The clause attempts to confer on an arbitrator powers which he would not otherwise have and aims to enable the same arbitrator to determine all the disputes.

Although in other JCT standard forms there is an obvious benefit to joining subcontractors, particularly named or nominated subcontractors, into any main contract arbitration proceedings over related issues, the same cannot be said of WCD 98. So it is difficult to see how, even if this provision could be operated in practice, it is relevant to this contract. To be effective in relation to 'domestic subcontract' a similar provision would, in any case, need to be inserted in those subcontracts and it is difficult to see how clause 39B.1.2 can effectively impose any such obligation on a domestic subcontractor if no such subcontract term exists. Nor, in like situations, could it be used to encompass any arbitration between employer and employer's agent, quantity surveyor, structural engineer or other professional consultant without their specific consent.

Most importantly, arbitration or litigation should be avoided if possible.

CHAPTER 14
SCOTTISH BUILDING CONTRACT WITH CONTRACTOR'S DESIGN

14.1 *The building contract*

The Scottish building contract incorporates CD 81, amends it and makes it suitable for use in Scotland. The latest revision is 1998. It is produced by the Scottish Building Contract Committee (SBCC) whose constituent bodies are:

- Royal Incorporation of Architects in Scotland
- Scottish Building Employers Federation
- Royal Institution of Chartered Surveyors in Scotland
- Convention of Scottish Local Authorities
- National Specialist Contractors Council
- Scottish Casec
- Association of Consulting Engineers (Scottish Group)
- Confederation of British Industry
- Association of Scottish Chambers of Commerce.

The document deals with CD 81 plus the 12 amendments current in August 1998. At the time of writing, a version to deal with WCD 98 had not been published although it is understood that work may be in hand. The amendments have not been adopted in their entirety and the contract contains a detailed list of the parts which are excluded.

The full title of the document is: Scottish Building Contract With Contractor's Design (August 1998 Revision) with Scottish Supplement 1982 (August 1998). The document is organised in five parts: the contract, the supplement forming appendix I, the abstract of conditions forming appendix II, alternative methods of payment forming appendix III, the Employer's Requirements, Contractor's Proposals, Contract Sum Analysis described in appendix IV and the SBCC correction sheet issued to JCT amendment 12: 1998.

The contract part is similar to the recitals and articles of CD 81.

The wording is slightly different, but the effect is the same overall. Clause 3 incorporates:

- Conditions of CD 81 and the VAT agreement
- JCT amendments 1 to 12 excluding all references to arbitration and certain other matters
- Abstract of conditions
- Alternative payment methods
- Employer's Requirements, Contractor's Proposals and Contract Sum Analysis as described in appendix IV.

The following matters are not included in the Scottish contract:

- Change of 'Works' to 'site' in appropriate circumstances
- Most of the revised determination clauses 27 and 28.

14.2 *Scottish supplement*

This includes detailed amendments to make CD 81 suitable for use in Scotland.

14.2.1 Interpretations and definitions

This is part of appendix I of the supplement. Clauses 1.1, 1.2 and 1.7 are deleted. In clause 1.3 a number of meanings of words or phrases are deleted and new meanings inserted. The most important of these are:

- Appendix 1, 2, or 3 defined as appendix II, III, or IV
- Arbitrator defined as arbiter
- Articles of agreement defined as the foregoing building contract.

A new clause 1.7 is added to contain definitions particular to Scotland:

- Execution of contract defined as formal adoption and signing
- Assignment defined as assignation
- Real or personal defined as heritable or moveable
- Section 117 of the Local Government Act 1972 defined as section 66 of the Local Government (Scotland) Act 1973 or any re-enactment thereof.

14.2.2 Clauses

Clauses 2.5.2 and 2.5.3, which deal with matters relating to the Defective Premises Act 1972, are deleted and a new clause 2.5.2 is inserted which reinstates the middle part of clause 2.5.3 to allow the contractor's liability for loss of use, loss of profit or other consequential loss to be limited to any amount noted in appendix II.

Clause 13 is amended slightly to make clear that the provisions referred to include without prejudice clause 30.10 which refers to the deduction of liquidate and ascertained damages.

Clause 15 has a phrase added which has the effect of extending the contractor's responsibility for loss or damage to materials or goods purchased before delivery to site under a separate contract provided for in clause 30.2B.4.

Clause 18.1.2 is deleted and the equivalent clause inserted to deal with Scottish terminology and usage, but to largely the same effect as the original clause in English law. Clauses 18.3.2.1 to 18.3.2.4, dealing with ownership of goods and materials, are also deleted.

Clause 27, determination by the employer, is deleted and a clause substituted which is to the same general effect. However, there is no express clause dealing with the employer's power to opt not to complete the Works.

Clause 28.1.3 rectifies the omission in the former CD 81 by inserting this clause to cover the employer's insolvency.

Amendments are made to clause 30 to comply with Scottish practice. Clause 30.2B.1.3, dealing with the value of materials, is deleted and a new clause 30.2B.4 is inserted to allow the employer to exclude the purchase of any materials or goods from the contract by entering into a separate contract for the purchase. The employer must have the contractor's consent before entering into a separate contract with a subcontractor, but the consent must not be unreasonably withheld. The contractor is still entitled to receive his cash discounts or other payments, but from the employer rather than the subcontractor.

There is provision for the parties to decide and indicate whether clauses 30.4.2.2 and 30.5.3.11 are to apply. These deal with the placing of retention into a separate bank account and the addition to the contract sum of the amount paid by the contractor under clauses 22B.2 or 22C.3 respectively. A note indicates that the clauses are not to apply if the employer is a local or a public authority. In the light of recent case law (see section 10.5) it is uncertain whether the provision will be effective in removing the contractor's right to have the retention, held as a trust, put into a separate account. Clauses

22B.2 and 22C.3 are to be deleted if the employer is a local authority, because most local authorities choose not to insure property.

A new clause 30.10 is inserted to make clear that clauses 30.5.3 and 30.8.1 (adjustment of the contract sum and the conclusivity of the final account and final statement respectively) do not prevent the employer from deducting or adding liquidate and ascertained damages to or from sums due one to the other.

14.2.3 Adjudication, arbitration and litigation

The whole of the arbitration clause 39 is to be deleted and new clauses 39A, 39B and 39C substituted.

The adjudication provisions are similar to 39A of WCD 98, but there are differences in arrangement and content. For example, either of the parties may request, and the adjudicator must provide, reasons for his decision, provided the request is made no later than seven days from the date his decision is delivered to both parties. It is also stressed by clause 39A.8.2 that, if he fails to give any contrary directions, the parties must comply with the adjudicator's decision immediately. Usefully, the adjudicator may issue directions about security for the principal sum.

So far as arbitration is concerned, if a dispute or difference arises (excluding disputes under clauses 31.9 or clause 3 of the VAT agreement), either party may serve notice on the other requesting concurrence in the appointment of an arbiter. If they fail to agree within 14 days, either party may request the appointor named in appendix II to name an arbiter. The arbiter's powers are set out in clause 39B.2. They are somewhat different to his powers in the unamended version. In the Scottish version his powers are:

- To direct whatever measurements and valuations may be desirable to determine the rights of the parties
- To ascertain and award any sum which should have been included in any payment
- To open up, review and revise any account, opinion, decision, requirement or notice
- To determine all matters in dispute as though no account, opinion, decision, etc. had been given; that is, provided they have been submitted to him for decision. Care must be exercised in drafting the notice of arbitration to include all matters upon which the arbiter's decision is required
- To award compensation or damages and expenses.

Clause 39B.3 makes the award final and binding but subject to the provisions of section 3 of the Administration of Justice (Scotland) Act 1972. Clause 39B.4 entitles the arbiter to appoint a clerk to assist him, to issue interim awards, to receive remuneration, to find the parties joint and severally liable for his fees and to dispense with a Deed of Submission.

In common with WCD 98, a litigation option is provided, termed 'Court Proceedings', if required instead of arbitration.

14.3 *Abstract of conditions*

This is the equivalent of appendix 1 of WCD 98; indeed it is very similar in content. It is appendix II of the Scottish supplement. The main differences are:

- No provision for electronic data interchange
- No reference to the Defective Premises Act 1972
- No reference to bonds for off-site materials
- The appointor of the arbiter is to be one of the following:
  - Chairman or Vice-chairman of the Scottish Building Contract Committee
  - Dean of the Faculty of Advocates
  - Sheriff of the Sheriffdom in which the Works are situated.

14.4 *Appendices III and IV*

Appendix III deals with alternative methods of payment. It is virtually the same as appendix 2 of WCD 98.

Appendix IV is identical to appendix 3 of WCD 98 except that there is no reference to article 4. This appendix allows space for the documents comprising the Employer's Requirements, the Contractor's Proposals and the Contract Sum Analysis.

TABLE OF CASES

The following abbreviations are used:

AC	Law Reports, Appeal Cases
App Cas	Law Reports, Appeal Cases
ALJR	Australian Law Journal Reports
All ER	All England Law Reports
BLISS	Building Law Information Subscriber Service
BLM	Building Law Monthly
BLR	Building Law Reports
C & P	Carrington & Payne Reports
Ch	Law Reports, Chancery Division
CILL	Construction Industry Law Letter
CLD	Construction Law Digest
Con LR	Construction Law Reports
Const LJ	Construction Law Journal
DLR	Dominion Law Reports
EG	Estates Gazette
Ex	Law Reports, Exchequer Division
KB	Law Reports, King's Bench Division
LGR	Local Government Reports
LT	Law Times reports
Lloyds Rep	Lloyds Law Reports
QB	Law Reports, Queen's Bench Division
TLR	The Times Law Reports
WLR	Weekly Law reports

CLAUSE INDEX

Clause Index

SUBJECT INDEX